国家林业和草原局普通高等教育"十三五"规划教材
高等院校林产化工专业系列教材

林产化工专业英语

Forest Products Chemical Processing Speciality English

李淑君　陈志俊　主编

中国林业出版社

图书在版编目（CIP）数据

林产化工专业英语 / 李淑君，陈志俊主编. —北京：中国林业出版社，2020.7
国家林业和草原局普通高等教育"十三五"规划教材　高等院校林产化工专业系列教材
ISBN 978-7-5219-0327-0

Ⅰ.①林…　Ⅱ.①李…　②陈…　Ⅲ.①林产工业–化学工业–英语–高等学校–教材
Ⅳ.①TQ35

中国版本图书馆 CIP 数据核字（2019）第 248298 号

中国林业出版社教育分社

策划编辑：杨长峰　吴卉　肖基浒　　　**责任编辑**：肖基浒
电　　话：(010)83143555　　　　　　　**传　　真**：(010)83143516

出版发行	中国林业出版社（100009　北京市西城区德内大街刘海胡同7号） E-mail: jiaocaipublic@163.com　电话：(010)83143520 http://www.forestry.gov.cn/lycb.html
经　销	新华书店
印　刷	三河市祥达印刷包装有限公司
版　次	2020年7月第1版
印　次	2020年7月第1次印刷
开　本	850mm×1168mm　1/16
印　张	8.75
字　数	344千字
定　价	30.00元

未经许可，不得以任何方式复制或抄袭本书之部分或全部内容。

版权所有　侵权必究

高等院校林产化工专业系列教材
编写指导委员会

主　任：王　飞

副主任：蒋建新　李淑君

委　员：(按姓氏笔画为序)

　　　　王　飞(南京林业大学)

　　　　王宗德(江西农业大学)

　　　　左宋林(南京林业大学)

　　　　李淑君(东北林业大学)

　　　　李湘洲(中南林业科技大学)

　　　　杨　静(西南林业大学)

　　　　黄　彪(福建农林大学)

　　　　蒋建新(北京林业大学)

《林产化工专业英语》编写人员

主　　编：李淑君　陈志俊
副 主 编：薛菁雯　李双明
编写人员：（按姓氏笔画排序）
　　　　　王永贵（东北林业大学）
　　　　　马春慧（东北林业大学）
　　　　　马艳丽（东北林业大学）
　　　　　任世学（东北林业大学）
　　　　　李　伟（东北林业大学）
　　　　　李双明（沈阳化工大学）
　　　　　李淑君（东北林业大学）
　　　　　陈志俊（东北林业大学）
　　　　　郭元茹（东北林业大学）
　　　　　薛菁雯（齐鲁工业大学）

序
PREFACE

随着地球上人口数量的不断增长和资源的不断消耗，人口、资源和环境之间的矛盾日益突出，生物质资源作为地球上唯一的可再生含碳资源，其开发利用受到越来越多的关注。据估计，以高聚糖和木质素为主要组分的植物生物质资源每年以约1600多亿吨的速度再生，其蕴含能量是石油年产量的15~20倍。因此，利用生物质资源生产人类所需要的燃料、化学品和材料等产品，满足人类日益增长的物质需要，成为现代世界科技和产业发展的主要方向和内容。

林产化学加工是以林业资源及其加工剩余物的加工利用为内涵的工程学科。该学科起源于20世纪上半叶由我国知名的林业教育家、科学家和社会活动家梁希先生在浙江大学和国立中央大学（现南京大学）所创立的森林利用化学研究室。在中华人民共和国成立初期，由于我国石油资源紧缺，石油和煤化工技术落后，依靠化学加工利用森林资源生产人们所必需的化学品，成为我国工业发展的重要组成部分，曾为我国的国民经济建设做出了重要贡献。在20世纪90年代，随着石油和煤等化石资源的大量开采、煤化工和石油化工产业的建立，我国林产化学加工领域受到影响，发展停滞不前。

在21世纪初期，由于对生物质资源加工利用的高度重视和快速发展，以林产生物质资源的化学加工利用为内涵的林产化学加工重新焕发出巨大的生机和活力，发展迅速。目前，传统的林产化学加工已发展成为化学和生物化学加工方法并重，化学、化工、生物、医学、材料等学科交融的现代林产化学加工学科，其产品范围包括化学品、能源与材料，服务领域涵盖现代农药、绿色食品、新能源和新材料等新兴产业和行业，在国民经济建设和社会发展中发挥着越来越重要的作用。

我国的林产化工专业由南京林业大学于1952年创立的，目前已在全国绝大部分的林业院校和部分农业院校设立了林产化工专业，为我国林产化学加工行业的建立和发展培养了大量的人才，是我国林产化学加工领域发展的基础。教材资源是专业建设和人才培养的核心内容和基础性工作，具有不可替代的作用。在20世纪80年代和90年代，以南京林业大学林产化工专业教师为主出版了第一套林产化工专业系列教材；此后，尽管部分教材进行了修订或编写，但总体上已不能反映林产化学加工工程学科发展的内涵变化和对人才培养的新要求，因此，教育部全国林业工程专业教学指导委员会林产化工分会主任单位南京林业大学组织召开了全国林产化工教材工作会议，召集全国林产化工领域一线的专家、教授规划编写能反映现代林产化学加工工程内涵的系列教材。系列教材不仅包括原有的《林产化学工艺学》《林产精细化学品工艺学》《林源生物活性物化学与利用》《林特产品化学与利用》等教材，还包括反映林产化学加工新方向的《生物质能源与化品》《生物资源生物化学加工利用》《生物基功能材料》《生物质热化学转化与炭材料》等教材，均列入国家林业和草原局"十三五"规划教材。

本套教材全面系统地反映了现代林产化工的知识技术体系，教材特色鲜明，充分体现了基础性、系统性和实践性，既重视学生知识技术体系的构建，又高度重视学生在学习过程中实践和创新能力的培养。它是全国林产化工专业建设和综合改革的主要成果之一，并为进一步打造高质量的林产化工精品课程奠定了坚实基础，必将促进全国林产化工专业建设的发展。本套教材不仅是林产化工专业学生系统学习林产化工知识技术的书籍，也是从事林产化工研究、教学与开发的研究生、教师以及科技工作人员全面学习掌握以林产资源为基础的生物质化学加工利用知识和技术的参考用书。

最后，借此机会感谢组织和参与本套教材编写的专家和学者，以及中国林业出版社对本套教材的编写和出版所付出的辛勤劳动和心血。

是为序。

中国工程院院士

2019 年 1 月

前　言
FOREWORD

　　林产化工与人类生活息息相关，是采用化学或者生物化学的手段对森林资源进行加工，来获得生物质能源、材料和化学品，致力于高效利用林业资源、实现可持续发展。

　　随着化石资源的日益短缺和人们环保意识的日益增强，森林作为陆地生态系统最大的主体和最重要的生物质资源，因其具有低碳循环、可再生、永续利用的突出优点，受到了全球的广泛关注。以森林资源为原料获得能源、材料和化学品也将在人类生产和生活中起到越来越重要的作用。具备林产化工专业知识技能并具有较高林产化工专业英语水平的专业人才越来越受到社会各界的欢迎。

　　林产化工专业英语是林产化工专业一门非常重要的课程，着重培养学生用英文学习和交流专业知识的能力。为了适应新时期对林产化工专业学生能力培养的要求，我们针对林产化工专业的学生，同时兼顾化学、化工等相关专业的学生，结合东北林业大学林产化工专业英语的教学实践，编写了本教材。本教材已列入国家林业和草原局普通高等教育"十三五"规划教材，也是高等院校林产化工专业系列教材之一。

　　本教材由东北林业大学李淑君、陈志俊主编，齐鲁工业大学薛菁雯和沈阳化工大学李双明任副主编，东北林业大学王永贵、李伟、马艳丽、马春慧、任世学、郭元茹参加编写。教材编写紧扣林产化工专业知识，从化学基础入手，涵盖了林产化工前沿领域，共42篇，每篇由课文、词汇表和难点注释、推荐阅读材料组成，选材全面，内容丰富，适宜林产化工及相关专业学生作为教材选用或自学参考。

　　由于编者水平有限，书中不妥之处在所难免，敬请读者批评指正。

编　者
2019 年 12 月

Contents

序 PREFACE
前言 FOREWORD

Lesson 1	Elements and Inorganic Compounds	(1)
Lesson 2	Kinds of Matter	(5)
Lesson 3	Atoms and Molecules	(7)
Lesson 4	Inorganic Compounds	(10)
Lesson 5	Organic Compounds and the Relative Nomenclature	(15)
Lesson 6	Biomass	(22)
Lesson 7	General Introduction of Wood	(25)
Lesson 8	A Brief Introduction of Lignocellulosic Biomass	(28)
Lesson 9	Lignocellulose Fractionation	(31)
Lesson 10	Cellulose: Structure and Properties	(33)
Lesson 11	Solution of Cellulose in Lonic Liquid	(36)
Lesson 12	Cellulose: Chemical Modification	(38)
Lesson 13	Cellulose Nanocrystals	(41)
Lesson 14	Cellulose Hydrogel	(43)
Lesson 15	Cellulose-based Aerogel Absorbers	(46)
Lesson 16	General Introduction of Lignin	(49)
Lesson 17	Characterization Techniques of Lignin (Ⅰ)	(52)
Lesson 18	Characterization Techniques of Lignin (Ⅱ)	(54)
Lesson 19	Depolymerization of Lignin	(57)
Lesson 20	Fast Pyrolysis of Lignin	(60)
Lesson 21	Association of Lignin	(62)
Lesson 22	Reactivity of Lignin in Lonic Liquids	(65)
Lesson 23	Functionalization of Lignin Hydroxyl Groups (Ⅰ)	(67)
Lesson 24	Functionalization of Lignin Hydroxyl Groups (Ⅱ)	(69)

Lesson 25	**Lignin-derived Polymers**	(72)
Lesson 26	**Biodegradation of Hemicellulose**	(74)
Lesson 27	**Tannins**	(76)
Lesson 28	**Flavonoids**	(79)
Lesson 29	**Modifications of Naturally Occurring Phenols**	(82)
Lesson 30	**Plant Oil**	(86)
Lesson 31	**Rosin**	(89)
Lesson 32	**Extraction Technology**	(92)
Lesson 33	**Hydrolysis**	(94)
Lesson 34	**Hydrothermal Treatment**	(96)
Lesson 35	**Active Carbon Materials**	(99)
Lesson 36	**Bioethanol**	(101)
Lesson 37	**Furfural**	(103)
Lesson 38	**Adhesives**	(106)
Lesson 39	**Alkaloids**	(108)
Lesson 40	**Biodiesel**	(111)
Lesson 41	**Biomass-derived Plastic Materials**	(114)
Lesson 42	**Biomass-derived Carbon Dots**	(117)
References		(121)

Lesson 1   Elements and Inorganic Compounds

1.1　Elements

　　Substances which cannot be broken down chemically into simpler substances have historically been known as elements. Chemical elements are symbolized by one-or two-letter abbreviations derived from their modern names, or in some cases from their old Latin names.

Table 1.1　Symbols for Some of the Elements

Modern name	Symbol	Modern name	Symbol	Modern name	Symbol	Modern name	Symbol
Hydrogen	H	Boron	B	Carbon	C	**Nitrogen**	N
Oxygen	O	**Magnesium**	Mg	Aluminum	Al	**Silicon**	Si
Phosphorus	P	**Sulfur**	S	**Chlorine**	Cl	Calcium	Ca
Titanium	Ti	Chromium	Cr	Manganese	Mn	Nickel	Ni
Zinc	Zn	**Bromine**	Br	**Iodine**	I	Barium	Ba

Table 1.2　Elements with Symbols Based on Their Old Latin Names

Modern name	Symbol	Derivation of Symbol	Modern name	Symbol	Derivation of Symbol
Antimony	Sb	Stibium	Mercury	Hg	Hydrargyrum
Copper	Cu	Cuprum	**Potassium**	K	Kalium
Gold	Au	Aurum	Silver	Ag	Argentum
Iron	Fe	Ferrum	**Sodium**	Na	Natrium
Lead	Pb	Plumbum	Tin	Sn	Stannum

1.2　Periodic table of the elements

　　The periodic table is a tabular arrangement of the chemical elements, ordered by their atomic number (number of protons in the nucleus), electron configurations, and recurring chemical properties. There are 118 elements on the periodic table. Of these 92 are naturally occurring. Every symbol of the elements is shown in one square, and there are also a couple of numbers in the square. The first is atomic number and the other is the atomic mass. The table also shows four rectangular blocks: s-, p-, d-, and f-block. In general, within one row (period) the elements are metals on the left side, and non-metals on the right side.

Table 1.3 Peeiodic Table of the Elements

1A																	8A
1 H 1.00794	2A											3A	4A	5A	6A	7A	2 He 4.002802
3 Li 6.941	4 Be 9.012182											5 B 10.811	6 C 12.0107	7 N 14.0067	8 O 15.9994	9 F 18.9984032	10 Ne 20.1797
11 Na 22.989789	12 Mg 24.3050	3B	4B	5B	6B	7B	—	8B	—	1B	2B	13 Al 28.9815388	14 Si 28.0855	15 P 30.973762	16 S 32.065	17 Cl 35.453	18 Ar 39.948
19 K 39.0983	20 Ca 40.078	21 Sc 44.955912	22 Ti 47.867	23 V 50.9415	24 Cr 51.9961	25 Mn 54.938045	26 Fe 55.845	27 Co 58.933195	28 Ni 58.6934	29 Cu 63.546	30 Zn 65.38	31 Ga 69.723	32 Ge 72.64	33 As 74.92160	34 Se 78.96	35 Br 79.904	36 Kr 83.798
37 Rb 85.4678	38 Sr 87.62	39 Y 44.90585	40 Zr 91.224	41 Nb 92.96638	42 Mo 95.98	43 Tc [98]	44 Ru 101.07	45 Rh 102.90550	46 Pd 106.42	47 Ag 107.8882	48 Cd 112.411	49 In 114.878	50 Sn 118.710	51 Sb 121.760	52 Te 127.60	53 I 126.90447	54 Xe 131.293
55 Cs 132.9054519	56 Ba 137.327	57-71 Lanthanides	72 Hf 178.49	73 Ta 180.94788	74 W 188.84	75 Re 188.207	76 Os 190.23	77 Ir 192.217	78 Pt 195.084	79 Ag 196.96669	80 Hg 200.59	81 Tl 204.3833	82 Pb 207.2	83 Bi 208.98040	84 Po [209]	85 At [210]	86 Rn [222]
87 Fr [223]	88 Ra [226]	89-103 Actinides	104 Rf [267]	105 Db [268]	106 Sg [271]	107 Bh [272]	108 Hs [270]	109 Mt [276]	110 Ds [281]	111 Rg [280]	112 Cn [285]	113 Uut [284]	114 Fl [289]	115 Uup [288]	116 Lv [293]	117 Uus [294]	118 Uuo [294]

Lanthanides	57 La 138.90547	58 Ce 140.116	59 Pr 140.90765	60 Nd 144.242	61 Pm [145]	62 Sm 150.36	63 Eu 151.964	64 Rd 157.25	65 Tb 158.92535	66 Dy 162.500	67 Ho 164.93032	68 Er 167.259	69 Tm 168.93421	70 Yb 173.054	71 Lu 174.9668
Actinides	89 Ac [227]	90 Th 232.03806	91 Pa 231.03588	92 U 238.02891	93 Np [237]	94 Pu [244]	95 Am [243]	96 Cm [247]	97 Bk [247]	98 Cf [251]	99 Es [252]	100 Fm [257]	101 Md [258]	102 No [259]	103 Lr [282]

The rows of the table are called periods; the columns are called groups. The groups (columns) have names as well as numbers: for example, Group 7A elements are the **halogens**; and group 8A, the noble gases. The periodic table can be used to derive relationships between the properties of the elements, and predict the properties of new elements yet to be discovered or synthesized. The periodic table provides a useful framework for analyzing chemical behavior, and is widely used in chemistry and other sciences.

Each chemical element has a unique atomic number representing the number of protons in its nucleus. Most elements have differing numbers of neutrons among different atoms, with these variants being referred to as isotopes. For example, carbon has three naturally occurring isotopes: all of its atoms have six protons and most have six neutrons as well, but about one percent have seven neutrons, and a very small fraction have eight neutrons. **Isotopes** are never separated in the periodic table; they are always grouped together under a single element. Elements with no stable isotopes have the atomic masses of their most stable isotopes, where such masses are shown, listed in parentheses.

In the standard periodic table, the elements are listed in order of increasing atomic number (the number of protons in the nucleus of an atom). A new row (period) is started when a new electron shell has its first electron. Columns (groups) are determined by the electron configuration of the atom; elements with the same number of electrons in a particular subshell fall into the same columns (e.g. oxygen and selenium are in the same column because they both have four electrons in the outermost p-subshell). Elements with similar chemical properties generally fall into the same group in the periodic table, although in the f-block, and to some respect in the d-block, the elements in the same period tend to have similar properties, as well. Thus, it is relatively easy to predict the chemical properties of an element if one knows the properties of the elements around it.

1.3　Elements of life

　　Life involves many chemical reactions. Chemical reactions occur constantly within itself and are essential to life. For example, the conversion of carbon dioxide and water to **glucose** using sunlight as an energy source by plants is a chemical reaction, and it is known as **photosynthesis**. Animals use chemical reactions to break down food sources, such as carbohydrates, fats, and proteins to release their energies to fuel cell processes.

　　In biology, living organisms are mainly composed of only four elements, i.e. oxygen, carbon, **hydrogen** and **nitrogen**. There are small amounts of other elements found in the living organisms, for example calcium which found in bones. Phosphorus and **magnesium** are also found in small amounts. There are other elements called trace elements, and these are also essential to life but they are found in very small or trace amounts. An example is iron. Iron is found in red blood cells. It is essential in allowing us to carry oxygen, using red blood cells. Iodine is another example. The thyroid gland needs iodine to function correctly. Without iodine, people will develop a condition called goiter.

　　Selected from: Gaoyuan Wei. *Introductory Chemistry Speciality English* (2nd Edition). Peking University Press, 2012.

　　https://en.wikipedia.org/wiki/Periodic_ table

Words and Expressions:

　　hydrogen　　['haɪdrədʒən]　氢，氢气
　　nitrogen　　['naɪtrədʒən]　氮，氮气
　　magnesium　　[mæg'niːziəm]　镁
　　silicon　　['sɪlɪkən]　硅
　　phosphorus　　['fɒsfərəs]　磷
　　sulfur　　[sʌlfə(r)]　n. 同"sulphur"，vt. 用硫黄处理
　　chlorine　　['klɔːriːn]　氯
　　bromine　　['brəʊmiːn]　溴
　　iodine　　['aɪədiːn]　碘
　　potassium　　[pə'tæsiəm]　钾
　　sodium　　['səʊdiəm]　钠
　　halogen　　['hælədʒən]　卤素
　　isotope　　['aɪsətəʊp]　同位素
　　proton　　['prəʊtɒn]　质子
　　glucose　　['gluːkəʊs]　葡萄糖
　　photosynthesis　　[ˌfəʊtəʊ'sɪnθəsɪs]　光合作用

Notes:

　　1) The periodic table is a tabular arrangement of the chemical elements, ordered by their atomic number (number of protons in the nucleus), electron configurations, and recurring chemical

properties.

元素周期表是根据元素的原子序数(原子核中的质子数)、电子排布和重复出现的化学性质将化学元素以表格的方式排序而成的列表。

2) Each chemical element has a unique atomic number representing the number of protons in its nucleus. Most elements have differing numbers of neutrons among different atoms, with these variants being referred to as isotopes.

每种化学元素都有一个唯一的原子序数，表示其原子核中的质子数。大多数元素在不同的原子中具有不同数量的中子，这些变体被称为同位素。

Recommended Reading Materials:

Brian T. Farrer, Vincent L. Pecoraro, in *Encyclopedia of Physical Science and Technology* (Third Edition), 2003.

Lesson 2 Kinds of Matter

2.1 Pure substances and mixtures

Physical properties are those that can be measured or observed without changing the identity or composition of a substance. Chemical properties can only be observed in chemical reactions, in which the identity of at least one substance is changed. A pure substance always has the same physical and chemical properties and is either an element or a compound. An element is a substance that contains only atoms of the same atomic number. (An atom can be defined as the smallest particle of an element that can participate in a chemical reaction.) A chemical compound is a substance in which atoms of two or more elements are combined in a definite ratio. A mixture contains two or more substances that retain their identities. Mixtures are divided into **homogeneous** mixtures and **heterogeneous** mixtures. The substances in a homogeneous mixture are thoroughly intermingled, and the composition and appearance of the mixture are uniform throughout, while a heterogeneous mixture is a mixture in which the individual **components** of the mixture remain physically separate and can be seen as separate components, although in some cases a microscope is needed. Any homogeneous mixture of two or more substances is a **solution**. The **solute**—the component present in the smaller amount—is said to be dissolved in the **solvent**. In an aqueous solution the solvent is water.

2.2 Elements, matter, and compounds

Elements are substances that cannot be broken down into simpler substances through chemical reactions. Matter is anything that take some space. Elements are a form of matter.

Compounds are composed of different elements that are combined in a fixed ratio. O_2 is two oxygen atoms bonded together. These are molecules. However, since these are the same type of atom, this is not a compound. H_2O or water is an example of a compound. This is two atoms of hydrogen bonded to one atom of oxygen. The fixed ratio and order in which the different elements combined is also very important. For example, water (H_2O) and hydrogen **peroxide** (H_2O_2) are both composed of hydrogen and oxygen. However, they have very different properties. Water is the solvent of life, or hydrogen peroxide has **bleaching**, **de-inking** and **oxidizing** ability.

2.3 States of matter

There are three common states of matter: gaseous, liquid, and solid. Transitions between these are known as changes of state. Not all substances can exist in all three states.

Selected from: Gaoyuan Wei. *Introductory chemistry speciality English*. 2nd Edition. Peking University Press, 2012.

Words and Expressions:

  homogeneous [ˌhɒməˈdʒiːniəs] *adj.* 由相同(或同类型)事物(或人)组成的
  heterogeneous [ˌhetərəˈdʒiːniəs] *adj.* 由很多种类组成的；各种各样的
  component [kəmˈpəʊnənt] *n.* 组成部分；成分；*adj.* 成分的；组成的；构成的
  solution [səˈluːʃn] *n.* 溶液
  solute [ˈsɒljuːt] *n.* 溶质；溶解物
  solvent [ˈsɒlvənt] *n.* 溶剂；溶媒；*adj.* 有溶解力的；可溶解的
  peroxide [pəˈrɒksaɪd] *n.* 过氧化物
  bleaching [ˈbliːtʃɪŋ] *v.* (使)漂白，退色；bleach 的现在分词
  de-inking [diːˈɪŋkɪŋ] *v.* (使)脱墨，除去污渍；de-ink 的现在分词
  oxidizing [ˈɒksɪdaɪzɪŋ] *v.* (使)氧化；(尤指使)生锈；oxidize 的现在分词

Notes:

1) Physical properties are those that can be measured or observed without changing the identity or composition of a substance. Chemical properties can only be observed in chemical reactions, in which the identity of at least one substance is changed.

物理性质是那些可以在不改变物质的特性或组成的情况下被测量或观察到的性质。化学性质只能在至少一种物质的特性改变的化学反应中观察到。

2) Mixtures are divided into homogeneous mixtures and heterogeneous mixtures. The substances in a homogeneous mixture are thoroughly intermingled, and the composition and appearance of the mixture are uniform throughout, while a heterogeneous mixture is a mixture in which the individual components of the mixture remain physically separate and can be seen as separate components, although in some cases a microscope is needed.

混合物分为均质混合物和非均质混合物。均质混合物中的物质完全混合，混合物的组成和外观始终是均一的；而非均质混合物是一种混合物，在这种混合物中，混合物的各个组分在物理上保持分离，可以被视为分开的组分，尽管在某些情况下需要显微镜。

Recommended Reading Materials:

  https://courses.lumenlearning.com/boundless-chemistry/chapter/classification-of-matter/

Lesson 3 Atoms and Molecules

3.1 Atoms

All matter is composed of tiny particles called atoms, which are themselves composed of smaller particles. An atom has a dense central core, or nucleus, containing positively charged protons and uncharged neutrons. Much lighter, negatively charged electrons occupy a relatively large space around the nucleus.

3.2 Molecules

A molecule is a combination of two or more atoms that are held together by chemical bonds. It's the smallest unit of a compound that still displays the properties associated with that compound. Molecules may contain two atoms of the same element, such as Oxygen (O_2), and Hydrogen (H_2). Or they may consist of two or more different atoms, such as **Carbon Tetrachloride** (CCl_4) and Water (H_2O). In the study of chemistry, molecules are usually discussed in term of their **molecular weights** and moles. Ionic compounds, such as **Sodium Chloride** (NaCl) and **Potassium Bromide** (KBr), do not form true molecules. And their solid state—this substances form three-dimensional array of charged particles. In such a case, molecular weight had no meaning, so the term **formula weight** is used instead.

3.3 Molecular weights and Formula weight

The molecular weight of a molecule is calculated by adding the atomic weights in Atomic Mass Units (or AMU) of the atoms in the molecule.

The formula weight of an ionic compound is calculated by adding its atomic weights according to its **empirical formula**.

3.4 Definition of a molecule

Molecules are formed when two or more atoms are held together by chemical bonds in a specific arrangement.

3.5 Molecules formed by covalent bonds

For example, the oxygen molecule consists of two atoms of oxygen. In the case of oxygen, the bond that holds the two atoms together is known as a **covalent bond**, and here is how it works.

The oxygen atom has 8 each of protons, neutrons, and electrons. The protons and neutrons are found in the center of the atom, known as the nucleus, and the electrons surround the nucleus in layers, or shells. The oxygen atom has 2 electrons in its first shell, and 6 in its second and outer shell. However, in chemistry, there exists the octet rule, which states that atoms generally strive to have 8 electrons in their outer shell. The oxygen therefore is 2 electrons away from a complete outer shell. When it binds with another oxygen atom, they can share two pairs of electrons and so each will have 8. When atoms share electrons like that, they form a molecule through a covalent bond.

3.6 Molecules formed by ionic bonds

Molecules can also be formed through other kinds of chemical bonds, for example, an ionic bond. An example this is sodium chloride, or table salt. In an ionic bond, one atom has a much greater ability to attract electrons than the other atom. In this case, the chlorine, which is only 1 electron short of a complete outer shell, will steal that electron from the sodium, which has only 1 outermost electron. This turns the sodium into a positively charged ion, and the chlorine into a negatively charged ion, and the two atoms end up being held together by an electrostatic charge.

3.7 Meaning of a mole

A mole is simply a counting number, much like a dozen. It is defined as the quantity of substance that has the same number of particles that are found in 12.000 grams of carbon-12. This number—Avogadro's number is 6.022 times ten to the power 23. The mass in gram of one mole of a compound is equal to the molecular weight of the compound in Atomic Mass Units. One mole of a compound contains 6.022 times ten to the power 23 molecules of the compound. The mass of one mole of a compound is called its molar weight or molar mass. The units for molar weight or molar mass are grams per mole. Here is the formula for determining the number of moles of a sample. Mole equals weight of a sample over molar weight.

3.8 Examples of finding molar mass

Moles make it easier to quantify chemical substances. The periodic table lists the weights of all the elements in grams per mole. For example, oxygen has molar mass of 16, which means 6.022×10 to the 23 atoms of oxygen weigh 16 grams. Hydrogen has a molar mass of 1. To find out the molar mass of a molecule, such as water, we need to add up the molar masses of the atoms it is composed of. Since water has two atoms of hydrogen and 1 atom of oxygen, its molar mass is 18 grams per mole.

Selected from: www.about.com

Words and Expressions:

  carbon tetrachloride [ˌkɑːbən ˈtetrəˈklɔːraɪd] *n.* 四氯化碳
  molecular weight [məˈlekjələ(r) weɪt] 分子量；相对分子质量
  sodium chloride [ˌsəʊdiəm ˈklɔːraɪd] *n.* 氯化钠；食盐
  potassium bromide [pəˈtæsiəm ˈbrəʊmaɪd] *n.* 溴化钾
  formula weight [ˈfɔːmjələ weɪt] [化学]式量；分子量
  empirical formula [ɪmˈpɪrɪkl ˈfɔːmjələ] 经验公式；实验式
  covalent bond [ˌkəʊˈveɪlənt bɒnd] 共价键

Notes:

1) All matter is composed of tiny particles called atoms, which are themselves composed of smaller particles. An atom has a dense central core, or nucleus, containing positively charged protons and uncharged neutrons. Much lighter, negatively charged electrons occupy a relatively large space around the nucleus.

所有物质都是由称为原子的微小粒子组成的，原子本身又是由更小的粒子组成的。原子有一个致密的中心核，或称原子核，包含带正电的质子和不带电的中子。更轻的带负电荷的电子在原子核周围占据了相对较大的空间。

2) It is defined as the quantity of substance that has the same number of particles that are found in 12.000 grams of carbon-12. This number—Avogadro's number is 6.022 times ten to the power 23. The mass in gram of one mole of a compound is equal to the molecular weight of the compound in Atomic Mass Units. One mole of a compound contains 6.022 times ten to the power 23 molecules of the compound.

它(摩尔)的定义是含有与 12.000 克 ^{12}C 中发现的粒子数量相同的物质的数量。这个数字——阿佛加德罗常数是 6.022 乘以 10 的 23 次方。以克为单位的 1 摩尔化合物的质量等于以原子质量单位表示的该化合物的分子量。1 摩尔化合物包含该化合物 6.022 乘以 10 的 23 次方个分子。

Recommended Reading Materials:

  https://study.com/academy/lesson/what-are-atoms-molecules-definition-differences.html

Lesson 4 Inorganic Compounds

4.1 Nomenclature for inorganic compounds

The rules that govern the naming of chemical compounds are known collectively as chemical **nomenclature**. Many inorganic compounds are ionic compounds, consisting of cations and anions joined by ionic bonding. In the Stock system, the name of a **cation** consists of the name of the element, the charge on the ion as a Roman numeral in parentheses, and the word "ion". The name of a monatomic anion consists of the name of the element with the ending "ide", followed by the word "ion". Examples of salts (which are ionic compounds) are magnesium chloride $MgCl_2$, which consists of magnesium cations Mg^{2+} and chloride anions Cl^-; or sodium oxide Na_2O, which consists of sodium cations Na^+ and oxide anions O^{2-}. The following prefixes are often used to indicate the number of atoms of each element present.

Table 4.1 Some prefixes indicated numbers

No. indicated	Prefix	Example of Compound	full name of the Compound
1	Mono-	CO	Carbon monoxide
2	Di-	CO_2	Carbon dioxide
3	Tri-	SO_3	Sulfur trioxide
4	Tetra-	$TiCl_4$	Titanium tetrachloride
5	Penta-	N_2O_5	Dinitrogen pentoxide

Salts are ionic compounds formed between cations and the anions of acids. In any salt, the proportions of the ions are such that the electric charges cancel out, so that the bulk compound is electrically neutral. The ions are described by their oxidation state and their ease of formation can be inferred from the **ionization potential** (for cations) or from the electron affinity (anions) of the parent elements.

Table 4.2 Names of some monatomic ions

Name of cations	Symbol	Name of anions	Symbol
Sodium ion	Na^+	Chloride ion	Cl^-
Iron (II) ion	Fe^{2+}	Fluoride ion	F^-
Iron (III) ion	Fe^{3+}	Bromide ion	Br^-
Copper (II) ion	Cu^{2+}	Iodide ion	I^-
Copper (I) ion	Cu^+	Sulfide ion	S^{2-}
Manganese (IV) ion	Mn^{4+}	Nitride ion	N^{3-}
Manganese (II) ion	Mn^{2+}	Hydride	H^-
Barium ion	Ba^{2+}	Phosphide ion	P^{3-}

Naming acid salt ions. The anions of acid salt are named with the word "hydrogen" placed before the name of the normal anion. HSO_4^- is the hydrogen **sulfate** ion. To denote two atoms, the prefix di-is used. HPO_4^{2-} is the hydrogen **phosphate** ion, while $H_2PO_4^-$ is the dihydrogen phosphate ion. In an older naming system, the prefix bi-was used instead of the word hydrogen when one of two hydrogen atoms were replaced. Thus, HCO_3^- was called the bicarbonate ion instead of the more modern name, hydrogen **carbonate** ion.

Table 4.3 Names of some polyatmic ions

Name of ions	Symbol	Name of ions	Symbol	Name of ions	Symbol
Ammonium ion	NH_4^+	Hydrogen sulfate ion	HSO_4^-	Hydroxide ion	OH^-
Hydroxonium	H_3O^+	Sulfite ion	SO_3^{2-}	Acetate ion	CH_3COO^-
Chromate ion	CrO_4^{2-}	Hydrogensulfite ion	HSO_3^-	Nitrate ion	NO_3^-
Dichromate ion	$Cr_2O_7^{2-}$	Hydrogen sulfide ion	HS^-	Nitrite ion	NO_2^-
Bromate ion	BrO_3^-	Carbonate ion	CO_3^{2-}	perchlorate ion	ClO_4^-
Iodate ion	IO_3^-	Hydrogen carbonate ion	HCO_3^-	Chlorate ion	ClO_3^-
Cyanide ion	CN^-	Hydrogen phosphate ion	HPO_4^{2-}	Chlorite ion	ClO_2^-
Thiocyanate ion	SCN^-	Dihydrogen phosphate ion	$H_2PO_4^-$	Hypochlorite ion	ClO^-
Sulfate ion	SO_4^{2-}	Hydrogen phosphite ion	HPO_3^{2-}	Phosphate ion	PO_4^{3-}

Naming inorganic acids and salts. The name of an inorganic acid can be obtained by replacing the "-ate" ending of an anion with "-ic acid" or replacing the "-ite" ending with "-ous acid." If the anion ends in-ide, add the prefix hydro-and change the ending to "-ic acid". Important classes of inorganic salts are the oxides, the carbonates, the sulfates and the **halides**. Many inorganic compounds are characterized by high melting points. Inorganic salts typically are poor conductors in the solid state. Other important features include their solubility in water and ease of **crystallization**. Where some salts (e.g., NaCl) are very soluble in water, others (e.g., SiO_2) are not.

Table 4.4 Names of some inorganic compounds

Name of compounds	Symbol	Name of compounds	Symbol
Hydrogen peroxide	H_2O_2	perchloric aid	$HClO_4$
Diphosphoric pentoxide	P_2O_5	Chloric acid	$HClO_3$
Dinitrogen oxide	N_2O	Chlorous acid	$HClO_2$
Calcium oxide	CaO	Hypochlorous acid	$HClO$
Hydrogen chloride	HCl	Boric acid	H_3BO_3
Hydrochloric acid	$HCl\,(aq)$	Oxalic acid	$H_2C_2O_4$
Sulfuric acid	H_2SO_4	Calcium oxalate	CaC_2O_4
Sulfurous acid	H_2SO_3	Carbon bisulfide	CS_2
Nitric acid	HNO_3	Ammonia	NH_3
Nitrous acid	HNO_2	Ammonium bicarbonate	NH_4HCO_3
Acetic acid	CH_3COOH	Magnesium hydroxide	$Mg(OH)_2$
Carbonic acid	H_2CO_3	Sodium chloride	$NaCl$
Phosphoric acid	H_3PO_4	Potassium permanganate	$KMnO_4$
Phosphorous acid	H_3PO_3	Copper (II) sulfate pentahydrate	$CuSO_4 \cdot 5H_2O$
Silicic acid	H_2SiO_3	Anhydrous copper (II) sulfate	$CuSO_4$ (without water)

Several types of chemical reactions

Synthesis. A **synthesis** reaction occurs when two or more elements or compounds combine and create a new and more complex product (A + B → C). An example of a synthesis reaction is the combination of sulfur and iron to form iron **sulfide**.

Decomposition. A **decomposition** reaction is one where a compound breaks down into smaller and less complex elements or compounds (AB → A + B or ABC → A + B + C). An example of a decomposition reaction is the electrolysis of water to make oxygen and hydrogen gas.

Single Displacement. A displacement reaction is a reaction where an element replaces, or displaces, a less active element in a compound. Iron will replace copper in a copper (II) chloride solution, producing iron (II) chloride and copper.

Double Displacement. A double displacement reaction is a reaction where two compounds react and form new compounds. The formation of more stable compounds, such as of a water or a **precipitate**, powers these reactions (AB + CD → AD + CB). Barium **nitrate** and sodium sulfate will form barium sulfate, the precipitate, and sodium nitrate.

Combustion. A combustion reaction is an **exothermic** reaction that occurs when oxygen and another element or compound are combined. A basic combustion reaction occurs when carbon in the form of charcoal reacts with oxygen to form carbon dioxide and heat.

The simplest inorganic reaction is double displacement when in mixing of two salts the ions are swapped without a change in oxidation state. In redox reactions one reactant, the oxidant, lowers its oxidation state and another reactant, the reductant, has its oxidation state increased. The net result is an exchange of electrons. Electron exchange can occur indirectly as well, e.g., in batteries, a key concept in electrochemistry.

When one reactant contains hydrogen atoms, a reaction can take place by exchanging protons in acid-base chemistry. In a more general definition, any chemical species capable of binding to electron pairs is called a Lewis acid; conversely any molecule that tends to donate an electron pair is referred to as a Lewis base. As a refinement of acid-base interactions, the HSAB (Hard Soft Acid Base) theory takes into account polarizability and size of ions.

Inorganic compounds are found in nature as minerals. Soil may contain iron sulfide as **pyrite** or calcium sulfate as **gypsum**. Inorganic compounds are also found multitasking as biomolecules: as **electrolytes** (sodium chloride), in energy storage (ATP) or in construction (the polyphosphate backbone in DNA).

The first important man-made inorganic compound was **ammonium** nitrate for soil fertilization through the Haber process. Inorganic compounds are synthesized for use as catalysts such as vanadium (V) oxide and titanium(III) chloride, or as reagents in organic chemistry such as lithium aluminium hydride.

Subdivisions of inorganic chemistry are organometallic chemistry, cluster chemistry and bioinorganic chemistry. These fields are active areas of research in inorganic chemistry, aimed toward new catalysts, superconductors, and therapies.

Selected from: Gaoyuan Wei. *Introductory Chemistry Speciality English*. 2nd Edition. Peking University Press, 2012.

www. about. com

Words and Expressions:
nomenclature　[nəˈmenklətʃə(r)]　符号表；命名法则；命名；命名法；标名
cation　[ˈkætaɪən]　n. 正离子；阳离子
ionization potential　[ˌaɪənaɪˈzeɪʃn pəˈtenʃl]　电离电位；电离电势
sulfate　[ˈsʌlfeɪt]　n. 硫酸盐；vt. 硫酸盐化用硫酸(盐)处理；使成硫酸盐
phosphate　[ˈfɒsfeɪt]　n. 磷酸盐；含磷化合物；磷肥
carbonate　[ˈkɑːbənət]　n. 碳酸盐；vt. 充二氧化碳于
halide　[ˈhæˌlaɪd]　n. 卤化物；卤素化合物；adj. 卤化物的；卤素的
crystallization　[krɪstəlaɪˈzeɪʃn]　n. 结晶化
synthesis　[ˈsɪnθəsɪs]　n. 综合；结合；综合体；合成
sulfite　[ˈsʌlˌfaɪt]　n. 亚硫酸根；亚硫酸盐
decomposition　[ˌdiːkɒmpəˈzɪʃn]　n. 腐烂；分解；
precipitate　[prɪˈsɪpɪteɪt, prɪˈsɪpɪtət]　n. 沉淀物；析出物
nitrate　[ˈnaɪtreɪt]　n. 硝酸盐
exothermic　[ˌeksəʊˈθɜːmɪk]　adj. 放热的
pyrite　[ˈpaɪraɪt]　n. 硫化铁；黄铁矿；硫铁矿；二硫化铁
gypsum　[ˈdʒɪpsəm]　n. 石膏
electrolyte　[ɪˈlektrəlaɪt]　n. 电解液；电解质
ammonium　[əˈməʊniəm]　n. 氨

Notes:

1) The simplest inorganic reaction is double displacement when in mixing of two salts the ions are swapped without a change in oxidation state. In redox reactions one reactant, the oxidant, lowers its oxidation state and another reactant, the reductant, has its oxidation state increased.

最简单的无机反应是置换反应。当两种盐混合时，离子发生交换而不改变氧化态。在氧化还原反应中，一种反应物(氧化剂)降低其氧化态，而另一种反应物(还原剂)则提高其氧化态。

2) When one reactant contains hydrogen atoms, a reaction can take place by exchanging protons in acid-base chemistry. In a more general definition, any chemical species capable of binding to electron pairs is called a Lewis acid; conversely any molecule that tends to donate an electron pair is referred to as a Lewis base. As a refinement of acid-base interactions, the HSAB (Hard Soft Acid Base) theory takes into account polarizability and size of ions.

在酸碱化学中，当一种反应物含有氢原子时，可以通过交换质子来进行反应。在更广义的定义中，任何能够与电子对结合的化学物质都被称为路易斯酸；相反，任何倾向于提供电子对的分子都被称为路易斯碱。作为酸碱相互作用的改进，HSAB(软硬酸碱)理论考虑了离子的极化率和大小。

3) Subdivisions of inorganic chemistry are organometallic chemistry, cluster chemistry and bioinorganic chemistry. These fields are active areas of research in inorganic chemistry, aimed toward

new catalysts, superconductors, and therapies.

无机化学细分为有机金属化学、团簇化学和生物无机化学。这些领域是无机化学研究的活跃领域，旨在开发新的催化剂、超导体和疗法。

Recommended Reading Materials:

https://www.britannica.com/science/inorganic-compound

Lesson 5   Organic Compounds and the Relative Nomenclature

5.1 Saturated and unsaturated hydrocarbons

Hydrocarbons are any of numerous organic compounds, such as methane and **benzene**, that contain only carbon and hydrogen.

5.1.1 Nomenclature for saturated hydrocarbons

Saturated aliphatic hydrocarbons (paraffin hydrocarbons) with the general formula C_nH_{2n+2} are called alkanes. The first four members of the alkane series are methane (CH_4), ethane (CH_3CH_3), propane ($CH_3CH_2CH_3$), butane ($CH_3CH_2CH_2CH_3$). For alkanes with more than four carbon atoms, a prefix (penta, C5; hexa, C6; hepta, C7; octa, C_8; nona, C_9; deca, C_{10}) is used to indicate the longest continuous chain of carbon atoms, and the suffix-ane is added to this prefix (one of the two "a"s is dropped), e. g., pentane, hexane.

The position and name of branches from the main chain, or of atoms other than hydrogen, are added as prefixes to the name of the longest hydrocarbon chain. The position of attachment to the longest continuous chain is given by a number obtained by numbering the longest chain from the end nearest the branch. In this way, the groups attached to the chain are designated by the lowest numbers, e. g., 2-chloropentane, 2-methylbutane, 2,2,4-trimethylpentane (or the common name, isooctane), 1-chloro-3,3-dimethylpentane, 2-methylpropane (isobutane).

Cycloalkanes (alicyclic hydrocarbons or cycloparaffin) have the general formula C_nH_{2n}. The nomenclature of cycloalkanes follows the same pattern used for the noncyclic alkanes. If there are substitutes, one is given the number 1 and others are given the lowest possible numbers, e. g., cyclopropane, 1,2-dichlorocyclohexane.

Carbon atoms can be classified as primary (the C joined to only one other carbon), secondary (to two other C's, as in the *sec*-butyl group), tertiary (as in the *tert*-butyl group), and quarternary carbon atom. Many hydrocarbons are unbranched, which are called normal hydrocarbons, e. g., n-hexane. As for groups of alkane, the -ane suffix of the relative alkane is replaced by -yl, such as methyl, ethyl, *n*-propyl, isopropyl, *n*-butyl, tert-butyl, *n*-pentyl, cyclopropyl, cyclobutyl, cyclohexyl. The names of some alkanes and relatives are listed in Table 5.1.

Table 5.1 Names of some alkanes and relatives

Catom number	Alkane	Prefix	Group	Alkene	Alkyne
1	Methane	Metha-	Methyl	Methene (:CH_2)	Methyne ($\equiv$CH)
2	Ethane	Etha-	Ethyl	Ethene	Ethyne
3	Propane	Propa-	Propyl	Propene	Propyne
4	Butane	Buta-	Butyl	Butene	Butyne
5	Pentane	Penta-	Pentyl	Pentene	Pentyne
6	Hexane	Hexa-	Hexyl	Hexene	Hexyne
7	Heptane	Hepta-	Heptyl	Heptene	Heptyne
8	Octane	Octa-	Octyl	Octene	Octyne
9	Nonane	Nona-	Nonyl	Nonene	Nonyne
10	Decane	Deca-	Decyl	Decene	Decyne

5.1.2 Nomenclature for unsaturated hydrocarbons

Basically, unsaturated hydrocarbons include alkenes (olefins, C_nH_{2n}), alkynes (acetylenes, C_nH_{2n-2}) and aromatic hydrocarbons.

Alkenes and alkynes. As shown in Table 5.1, to derive the systematic names of the individual alkenes, the -ane of the corresponding saturated hydrocarbon name is dropped and -ene is added if one double bond is present, adiene is added if two double bonds are present, and so on, e.g., ethene, 1,3-butadiene. Similarly, the individual alkynes are named by dropping the -ane and adding -yne, adiyne, atriyne, and so on. In either case, the position of the multiple bond is indicated by numbering from the end of the chain, starting at the end that will assign the lower number to the first carbon atom of the multiple bond. e.g., 1,3-pentadiyne.

In the common system of nomenclature, the -ane ending of the saturated hydrocarbon name is replaced by -ylene for the olefins. Compounds containing triple bonds are sometimes named as substituted acetylenes, (Because of the triple bond between the carbon atoms in acetylene, each carbon can have only one group attached to it.) e.g., ethylene (ethene, vinyl), propylene (propene, allyl), α-butylene (1-butene), isobutylene (2-methylpropene) β-butylene (2-butene, crotyl); acetylene (ethyne, ethynyl), methylacetylene (propyne, propargyl), vinylacetylene (1-buten-3-yne). Cycloalkenes are common, but cycloalkynes exist only for C_8 or larger rings. The triple bond is not flexible enough to fit easily into smaller rings. Beginning with propylene, alkenes of the homologous series of molecular formula C_nH_{2n} are isomeric with cycloalkanes, e.g., propene and cyclopropane. Acetylenes of the series of molecular formula C_nH_{2n-2} with n > 2 are isomeric with cycloalkenes, e.g., propyne and cyclo-propene.

Aromatic hydrocarbons and polycyclic aromatic hydrocarbons. In aromatic hydrocarbons derived from benzene, the added groups are referred to as substituents. The name of a single substituent is added to "benzene" as a prefix, as in ethylbenzene. Three structurally isomeric forms are possible for a disubstituted benzene, whether or not the substituents are the same. The three possibilities are designated ortho (abbreviated o-), meta (abbreviated m-), and para (abbreviated p-) as may be seen in the **xylenes**, e.g., m-xylene [$C_6H_4(CH_3)_2$].

Numbers are also used to show the positions of substituents in aromatic compounds. Unless there is no question of what the structure is, as in hexachlorobenzene, numbers are always used to locate the substituents when three or more are present in the ring.

The benzene molecule less one hydrogen atom is known as the phenyl group, C_6H_5. Diphenylmethane, for example, is $(C_6H_5)_2CH_2$.

Polycyclic aromatic hydrocarbons contain two or more aromatic rings fused together. ("Fused" rings have in common a bond between the same two atoms.) Some examples are: naphthalene, 1,4-dimethylnaphthalene, anthracene, and phenanthrene. Several resonance forms can be written for each of these compounds. Numbers are assigned by convention to carbon atoms in fused ring systems (except to those at the points of fusion, where substitution is not possible). The substituent locations are identified by the assigned numbers.

5.1.3 Nomenclature for functional groups

Organic compounds often contain one or more functional groups. Names of some common functional groups are given in Table 5.2.

Table 5.2 Names of some alkyl or aryl halides (RX or ArX)

Functional groups	Name	Functional groups	Name
-X	halo	-CHO	aldehyde or formyl
-OH	hydroxyl	-CO-	carbonyl or keto or oxo
-OR	alkoxy	-COOH	carboxyl
$-NR_3$	amino	$-CONH_2$	amido
R_nNH_{3-n}	amine	-COX	carbonyl halide
(n=1)	primary amine	-COOCOR	anhydride
(n=2)	secondary	-COOR	ester
(n=3)	tertiary	$R-SO_3H$	sulfonic acid
$-NO_2$	nitro	-CN	cyano
$R-NO_2$	nitroalkane	R-CN	nitrile

5.2 Functional groups with covalent single bonds

Alkyl and aryl halides (RX, ArX), alcohols (ROH), **phenols** (ArOH), **ethers** (ROR, where the R's may be the same or different), and **amines** (primary, RNH_2 and secondary, R_2NH, or tertiary, R_3N, in which the R's may be the same of different) contain the following functional groups in which there are only single bonds: -X, -OH, -OR, and NH_2, -NHR, $-NR_2$. Some names of some alkyl or aryl halides are listed in Table 5.3.

Table 5.3 Names of some alkyl or aryl halides

Compound	Name	Compound	Name
CH_3Br	methyl bromide (bromomethane)	CH_3I	methyl iodide (iodomethane)
CH_3CH_2Br	ethyl bromide (bromoethane)	$CH_2=CHCl$	vinyl chloride ompound (chloroethene)
$CH_3CH_2CH_2Br$	propyl bromide (1-bromopropane)	C_6H_5Cl	Chlorobenzene

Alcohols are classified as primary, secondary, or tertiary depending on whether the carbon atom to which the OH group is attached is primary, secondary, or tertiary. In the IUPAC system of nomenclature, the name is derived from the longest hydrocarbon chain that includes the OH group by dropping the final -e and adding -ol, as in methanol, ethanol, and cyclohexanol. When necessary, a number is used to show the position of the OH group. Numbering starts at the end of the chain nearest to the OH group. Names of some alcohols and phenols are listed in Table 5.4. Alcohols exhibit hydrogen bonding and those with low molecular masses are miscible with water. Phenols are weak acids. Ethers tend to be un-reactive, and low molecular mass ethers are often used as solvents. The amines are organic bases. Names of some ethers and amines are listed in Table 5.5.

Table 5.4 Names of some alcohols and phenols

Compound	Name	Compound	Name
CH_3OH	methyl alcohol (methanol)	$RO^- M^+$	alkoxide
CH_3CH_2OH	ethyl alcohol (ethanol)	CH_3CH_2ONa	sodium ethoxide
$CH_3CH(OH)CH_3$	isopropyl alcohol (2-propanol)	C_6H_5OH	phenol
$CH_3CH(OH)CH_2CH_3$	2-butanol	$CH_3C_6H_4OH$	p-methylphenol (p-cresol)
HO-cyclopentane	cyclopentanol	naphthol-OH	α-naphthol (1-naphthol)
tert-butyl structure	tert-butyl alcohol	benzene with two OH (ortho)	catechol
HO-CH2-CH2-OH	ethylene glycol (1,2-ethanediol)	HO-C6H4-OH (para)	hydroquinone
HO-CH2-CH(OH)-CH2-OH	glycerol (glycerin, 1,2,3-propanetriol)	HO-CH2-CH2-C6H5	β-phenylethyl alcohol (2-phenyl ethanol)

Table 5.5 Names of some ethers or amines

Compound	Name	Compound	Name
CH_3OCH_3	dimethyl ether or methyl ether (methoxymethane)	$CH_3CH_2OCH_2CH_3$	diethyl ether (ethoxyethane)
$CH_3OCH_2CH_2OCH_3$	ethylene glycol dimethyl ether ("glyme") (1,2-dimethoxyethane)	epoxide (triangle with O)	ethylene oxide (oxirane)
		six-membered ring with two O	dioxane (1,4-dioxin)
CH_3NH_2	methylamine	$(CH_5)_2NH$	dimethylamine
$(CH_3)_3N$	trimethylamine	$C_6H_5NH_2$	aniline
$(CH_3)_4NI$	Tetramethyl-ammonium iodide	pyridine ring	pyridine

5.3 Functional Groups with Covalent Double Bonds

Aldehydes (RCHO) and **ketones** (RCOR) (in which the R's may be the same or different) have carbonyl (-CO-) -containing **functional group**. A secondary alcohol can be oxidized to a ketone. Names of some aldehydes and ketones are listed in Table 5.6. A primary alcohol can be oxidized to an aldehyde, and further oxidation yields a **carboxylic acid** (RCOOH) which contains the **carboxyl group** -COOH. Carboxylic acids are generally weak acids and form salts when treated with bases. Names of some carboxylic acids and salts are listed in Table 5.7.

Table 5.6   Names of some aldehydes (RCHO) and ketones (RCOR')

Compound	Name	Compound	Name
HCHO	formaldehyde (methanal)	H_3C-CHO	acetaldehyde (ethanal)
O=CH—CH=CH_2	acrolein (propenal)	C_6H_5-CHO	benzaldehyde
CH_3—CO—CH_3	acetone (dimethyl ketone, 2-propanone)	CH_3-CO-CO-CH_3	biacetyl (dimethylglyoxal, 2,3-butane-dione)
H_3C-CO-$CH_2CH_2CH_3$	methyl n-propyl ketone (2-pentanone)	C_6H_5-CO-CH_3	methyl phenyl ketone (acetophenone)

Table 5.7   Names of some carboxylic acids (RCOOH) and salts

Compound	Name	Compound	Name
HCOOH	formic acid (methanoic acid)	CH_3COOH	acetic acid (ethanoic acid)
CH_2=CHCOOH	acrylic acid (propenoic acid)	$CH_3CHOHCOOH$	lactic acid
C_6H_5COOH	benzoic acid	o-$C_6H_4(COOH)_2$	phthalic acid
p-$C_6H_4(COOH)_2$	terephthalic acid	CH_3COOK	potassium acetate
CH_3CH_2COOH	propionic acid (propanoic acid)	$Ca(OOC-COO)$	calcium oxalate
CH_2COOH-$HOC(COOH)$-CH_2COOH	citric acid	NaOOC-CHOH-CHOH-COONa	sodium tartrate

Esters (RCOOR), **acyl** halides (RCOX), acid **anhydrides** (RCOOCOR), and **amides** ($RCONH_2$ and the N-substituted amides RCONHR and $RCONR_2$) are common types of organic compounds which have carbonyl-containing functional groups that are derived from carboxyl groups. In the two-word name for an ester, the first word is the name of the R' group in RCOOR'. This word could be methyl, ethyl, phenyl, or the like. The second word of the name is the name of the carboxylic acid with the final -ic replaced by -ate, identical with the name of the carboxylic anion. Names of some esters are listed in Table 5.8.

Table 5.8 Names of some esters

Compound	Name	Compound	Name
$HCOOCH_3$	methyl formate (methyl methanoate)	$CH_3COOC_2H_5$	ethyl acetate (ethyl ethanoate)
$C_6H_5COOCH_3$	methyl benzoate	$CH_2\!=\!CHCOOCH_3$	methyl acrylate (methyl propenoate)
(cyclic structure)	caprolactone	$o\text{-}C_6H_4(COOCH_2CH_3)_2$	diethyl phthalate

An acyl group (RCO-) is named by dropping the final -ic from the name of the corresponding carboxylic acid, and adding -yl. The complete name of the acyl halide is then the name of the acyl group, followed by the anionic name for the halogen atom. Anhydrides are given the name of the corresponding acid, followed by the word anhydride. Mixed anhydrides, from two different carboxylic acids, and cyclic anhydrides, are named similarly. Names of some acyl halides (RCOX) and carboxylic acid anhydrides (RCOOCOR') are listed in Table 5.9.

Table 5.9 Names of some acyl halides and carboxylic acid anhydrides

Compound	Name	Compound	Name
CH_3COCl	acetyl chloride (ethanoyl chloride)	$(CH_3CO)O(COC_3H_7)$	acetic butyric anhydride
C_6H_5COCl	benzoyl chloride	$(CH_3CO)_2O$	acetic anhydride (ethanoic anhydride)
$(C_6H_5CO)_2O$	benzoic anhydride	$o\text{-}C_6H_4(CO)_2O$	phthalic anhydride

The common name for an amide ($RCONH_2$) is derived by replacing the -yl of the aryl name by amide. Thus, acetamide is the name for CH_3CONH_2. Since the acyl group is named by replacing the -ic of the carboxylic acid name by -yl, one can equally well derive the amide name from the corresponding acid name by dropping the -ic (or -oic of the IUPAC name) and adding amide. If one or both of the hydrogen atoms on the amide nitrogen atom are replaced by hydrocarbon groups, the structure is named as an N-substituted amide.

Table 5.10 Names of some amides

Compound	Name	Compound	Name
$C_6H_5CONH_2$	benzamide	(structure)	N,N-dimethyl-formamide
(structure)	sulfanilamide	(structure)	N-phenylacetamide (acetanilide)

Selected from: Gaoyuan Wei. *Introductory Chemistry Speciality English* (2nd Edition). Peking University Press, 2012.

Words and Expressions:

hydrocarbon　[ˌhaɪdrəˈkɑːbən]　n. 烃；碳氢化合物
benzene　[ˈbenziːn]　n. 苯
Xylene　[ˈzaɪliːn]　n. 二甲苯
phenol　[ˈfiːnɒl]　n. 酚；苯酚；石碳酸
ether　[ˈiːθə(r)]　n. 醚；乙醚
amine　[əˈmiːn]　n. 胺
aldehyde　[ˈældɪhaɪd]　n. 醛；乙醛
ketone　[ˈkiːtəʊn]　n. 酮
carbonyl　[ˈkɑːbənɪl]　n. 羰基；金属羰基合物；adj. 含羰基的
functional group　[ˈfʌŋkʃənl gruːp]　官能团
carboxylic acid　[kɑːbɒkˈsɪlɪk ˈæsɪd]　羧酸
carboxyl group　[kɑːˈbɒksɪl gruːp]　羧基
ester　[ˈestə]　n. 酯
acyl halide　[ˈæsɪl ˈhælaɪd]　n. 酰卤
anhydride　[ænˈhaɪˌdraɪd]　n. 无水物；酐
amide　[ˈæmaɪd]　n. 氨基化合物；酰胺

Notes:

1) Carbon atoms can be classified as primary (the C joined to only one other carbon), secondary (to two other C:s, as in the sec-butyl group), tertiary (as in the tert-butyl group), and quarternary carbon atom.

碳原子可分为伯碳（C只与另一个碳连接）、仲碳（与另外两个C连接，如仲丁基）、叔碳（如叔丁基碳）和季碳原子。

2) Polycyclic aromatic hydrocarbons contain two or more aromatic rings fused together. ("Fused" rings have in common a bond between the same two atoms.) Some examples are: naphthalene, 1,4-dimethylnaphthalene, anthracene, and phenanthrene.

多环芳烃含有两个或两个以上稠合在一起的芳环。（"熔合"环在相同的两个原子之间有共同的键。）例如：萘、1,4-二甲基萘、蒽和菲。

Recommended Reading Materials:

https://study.com/academy/lesson/relative-configuration-in-organic-chemistry-definition-examples.html

Lesson 6　Biomass

Biomass is the plant material **derived** from the reaction between CO_2 in the air, water and sunlight, via **photosynthesis**, to produce **carbohydrates** that form the building blocks of biomass (Figure 6.1). Typically photosynthesis converts less than 1% of the available sunlight to store chemical energy. The solar energy driving photosynthesis is stored in the chemical bonds of the structural **components** of biomass. If biomass is processed efficiently, either chemically or biologically, by extracting the energy stored in the chemical bonds and the subsequent "energy" product combined with oxygen, the carbon is oxidised to produce CO_2 and water. The process is cyclical, as the CO_2 is then available to produce new biomass.

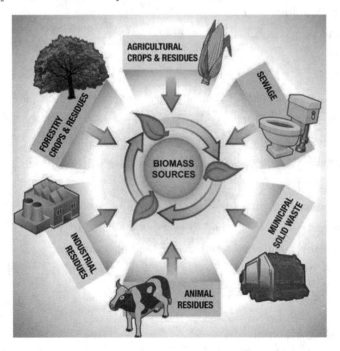

Figure 6.1　Classification of biomass resources

The value of a particular type of biomass depends on the chemical and physical properties of the large molecules from which it is made. Man for millennia has exploited the energy stored in these chemical bonds, by burning biomass as a fuel and by eating plants for the nutritional content of their sugar and **starch**. More recently, **fossilised** biomass has been **exploited** as coal and oil. However, since it takes millions of years to convert biomass into fossil fuels, these are not renewable within a time-scale mankind can use. Burning fossil fuels uses "old" biomass and converts it into "new" CO_2; which contributes to the "greenhouse" effect and depletes a non-renewable resource. Burning new biomass contributes no new carbon dioxide to the atmosphere, because replanting harvested biomass ensures that CO_2 is absorbed and returned for a cycle of new growth.

One important factor which is often overlooked when considering the use of biomass to **assist** in alleviating global warming, is the time lag between the instantaneous release of CO_2 from burning fossil fuels and its eventual **uptake** as biomass, which can take many years. One of the dilemmas facing the developed world is the need to recognize this time delay and take appropriate action to mitigate against the lag period. An equal dilemma faces the developing world as it consumes its biomass resources for fuel but does not implement a programme of replacement planting.

Numerous crops have been proposed or are being tested for commercial energy farming. Potential energy crops include woody crops and grasses/herbaceous plants (all perennial crops), starch and sugar crops and oilseeds. In general, the characteristics of the ideal energy crop are:

  i. high yield (maximum production of dry matter per hectare),
  ii. low energy input to produce,
  iii. low cost,
  iv. composition with the least contaminants,
  v. low nutrient requirements.

Desired characteristics will also depend on local climate and soil conditions. Water **consumption** can be a major constraint in many areas of the world and makes the drought resistance of the crop an important factor. Other important characteristics are pest resistance and **fertiliser** requirements.

Selected from: *Bioresource Technology*, 2002, 83: 37-46.

Words and Expressions:

biomass ['baɪə(ʊ)mæs] n. （单位面积或体积内的）[生态] 生物量；生物质
derive [dɪ'raɪv] vt. 源于；得自；获得；vi. 起源
photosynthesis [ˌfəʊtə(ʊ)'sɪnθɪsɪs] n. 光合作用
carbohydrate [kɑːbə'haɪdreɪt] n. [有化] 碳水化合物；[有化] 糖类
component [kəm'pəʊnənt] adj. 组成的，构成的；n. 成分；组件；[电子] 元件
starch [stɑːtʃ] n. 淀粉
fossilise ['fɒsɪlaɪz] vi. 变成化石；vt. 使成化石
exploit ['eksplɔɪt; ɪk'splɔɪt] vt. 开发，开拓；剥削；开采
assist [ə'sɪst] vt. 帮助；促进；n. 帮助；助攻
uptake ['ʌpteɪk] n. 摄取；领会；举起
consumption [kən'sʌm(p)ʃ(ə)n] n. 消费；消耗化肥
fertiliser ['fɜːtɪlaɪsə] n. 化肥

Notes:

1) If biomass is processed efficiently, either chemically or biologically, by extracting the energy stored in the chemical bonds and the subsequent 'energy' product combined with oxygen, the carbon is oxidised to produce CO_2 and water.

假如对生物质进行有效的化学或生物处理，提取其储存在化学键中的能量以及后续与氧

结合的"能量"产物,生物中的碳则会被氧化生成二氧化碳和水。

2) However, since it takes millions of years to convert biomass into fossil fuels, these are not renewable within a time-scale mankind can use.

然而,由于将生物质能转化为化石燃料需要数百万年的时间,因此,在人类能够使用的时间范围内,这些生物质能是不可再生的。

Recommended Reading Materials:

1. Zhang Z, Yang S, Li H, Zan Y, Li X, Zhu Y, Dou M, Wang F. Sustainable carbonaceous materials derived from biomass as metal-free electrocatalysts. *Adv. Mater.*, 2018, 1805718.

2. Wang Z, Smith A T, Wang W, Sun L. Versatile nanostructures from rice husk biomass for energy applications. *Angew. Chem. Int. Ed.*, 2018, 57: 13722-13734.

Lesson 7　General Introduction of Wood

Wood is a **biopolymer composite** that exhibits structural complexity spanning many orders of **magnitude** in length. Natural **diversity** further imparts **substantial variability** between the structures of wood for different species. The hierarchical structure of wood is visualized in Figure 7.1. At the **macroorganism** scale, interspecies differences in wood structure manifest as overall size and shape of the tree, **branching** patterns and frequencies, and bulk materials properties, such as density, thermal conductivity, strength, and elastic properties. The visually observable "wood grain", which results from the higher order arrangement and directionality of various cell types, also varies widely in appearance between species and contributes to the decorative value of the wood. Microscopic investigation of woody tissue reveals the highly ordered, interconnected network of pores formed by the cells during the growth of the tree. Interspecies differences in this microstructure are largely responsible for observed variations in aforementioned bulk thermal and mechanical properties.

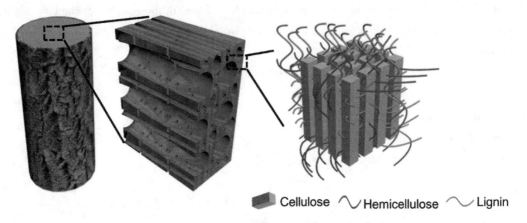

Figure 7.1　Hierarchical structure of wood

Wood cells generally resemble high-aspect ratio **cylinders** that run parallel to the axis of the **trunk** (axial tracheids and vessel cells) as well as cells that run radially from the heartwood to the bark (ray cells) to form a continuous network to transport and store water and nutrients throughout the organism. Several significant differences exist between the microstructure of hardwood and softwood species. First, the axial **tracheid** cells, which are commonly termed "fiber cells" or simply "fibers" in softwoods by the pulp and paper industry, range from 3 to 5 mm in length, whereas those cells in hard woods are typically shorter and range from 0.75 to 1.5mm in length. Second, hardwoods contain vessel elements, which are an additional class of axially **aligned** cells with a significantly larger diameter than the axial tracheids. Both hardwood and softwood cells contain **pits** that connect adjacent cells. Wood cell walls generally comprise three ultrastructural domains: the middle **lamella**, the primary wall, and the secondary wall. The middle lamella acts to adhere adjacent cells and is heavily lignified. In many cases, it is difficult to precisely delineate the primary cell wall from the middle lamella; thus, these two regions are often collectively termed the "compound middle lamella". The secondary wall contains the majority of the mass of wood and is typically composed of three distinct layers, denoted S1, S2, and S3.

The nanostructure of wood cell walls is a robust assembly of three primary biopolymers: cellulose, which is present as semicrystalline elementary **fibrils**; hemicellulose, which acts to decorate and cross-link elementary fibrils and facilitate interactions between other biopolymers; and lignin, which is an amorphous polymer of phenylpropanoids that associates closely with hemicellulose. The fully hemicellulose-sheathed elementary fibrils are termed microfibrils. Microfibrils and higher order bundles there of often termed "macrofibrils" form networks to physically support the cell wall. The molecular structure of each of these biopolymers will be described in greater detail in the following sections. Much work has been devoted to determining the precise configuration of various biopolymers within plant cell walls. Investigations of the primary cell wall have attracted much attention from the plant biology community because of its importance to cell wall biosynthesis, and this effort has provided generally accepted models for primary wall nanostructure. Furthermore, secondary cell walls have been the focus of the forest products industry because of their importance to engineering applications of wood. Similarly, the biofuels community has also turned their attention to secondary cell walls, because the majority of the biomass by weight is found in these walls. Secondary cell walls are also more recalcitrant to biochemical conversion processes, probably due to the increased lignin content in these walls.

Selected from: *Chem. Rev.* 2016, 116: 9305-9374.

Words and Expressions:

biopolymer composite [baɪəʊˈpɒlɪmə(r)] [ˈkɒmpəzɪt] 复合生物质高聚物
magnitude [ˈmæɡnɪtjuːd] n. 大小；量级
diversity [daɪˈvɜːsɪtɪ; dɪ-] n. 多样性；差异
substantial variability [səbˈstænʃ(ə)l] [ˌveərɪəˈbɪlətɪ] 实质性的变化
macroorganism [ˌmækrəʊˈɔːɡənɪzəm] 大生物体
branch [brɑːn(t)ʃ] v. 分支；出现分歧；n. 树枝，分枝；分部；支流
cylinder [ˈsɪlɪndə] n. 圆筒；汽缸；[数] 柱面；圆柱状物
trunk [trʌŋk] n. 树干；躯干
tracheid [ˈtreɪkiːd] n. [植] 管胞
align [əˈlaɪn] vt. 使结盟；使成一行；匹配；vi. 排列；排成一行
pit [pɪt] n. 深坑；陷阱；（物体或人体表面上的）凹陷；（木材）纹孔
lignify [ˈlɪɡnə, faɪ] vi. 木质化；vt. 使木质化
lamella [ləˈmelə] n. 薄板；薄片；薄层
fibril [ˈfaɪbrɪl] n. [生物] 纤丝，[生物] 原纤维；须根

Notes:

1) The visually observable "wood grain", which results from the higher order arrangement and directionality of various cell types, also varies widely in appearance between species and contributes to the decorative value of the wood.

源自于不同类型的木材细胞高阶有序排列与趋向性,木材呈现出"木纹"特征。这种木纹特征的表观随着其种类的变化而变化,并且对其装饰性有着重大的贡献。

2) First, the axial tracheid cells, which are commonly termed "fiber cells" or simply "fibers" in softwoods by the pulp and paper industry, range from 3 to 5 mm in length, whereas those cells in hard woods are typically shorter and range from 0.75 to 1.5mm in length.

首先,通常被制浆造纸行业称为"纤维细胞"或简称为"纤维"的轴向管胞细胞,其长度在 3~5 mm 之间,而硬木的管胞细胞通常较短,长度在 0.75~1.5 mm 之间。

Recommended Reading Materials:

1. Jiang F, Li T, Li Y, Zhang Y, Gong A, Dai J, Hitz E, Luo W, Hu L. Wood-based nanotechnologies toward sustainability. *Adv. Mater.*, 2018, 30, 1703453.

2. Zhu M, Song J, Li T, Gong A, Wang Y, Dai J, Yao Y, Luo W, Henderson D, Hu L. Highly anisotropic, highly transparent wood composites. *Adv. Mater.*, 2016, 28: 5181-5187.

Lesson 8　A Brief Introduction of Lignocellulosic Biomass

Lignocellulose is the major structural component of plants, and is by far the most **abundant** type of terrestrial biomass. It is a composite material mainly consisting of cellulose (40% ~ 60%), hemicellulose (10% ~ 40%), and lignin (15% ~ 30%); and can be found in both woody (e. g. pine, poplar, birch) and **herbaceous** biomass (e. g. switchgrass, miscanthus, corn stover) (Figure 8.1). The cellulose portion is exclusively composed of **glucose** units, which are linked in a linear fashion via β-1, 4-glycosidic bonds. The resulting polymer chains can have a polymerisation degree of up to 10 000 units. These chains interact with each other via hydrogen bonds and **van der Waals forces**, eventually giving rise to rigid, semi-crystalline cellulose fibres. Due to these strong interaction forces, cellulose fibres are insoluble in most conventional solvents, including water.

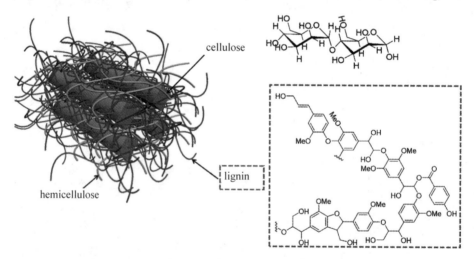

Figure 8.1　Chemical components of lignocellulosic biomass

The second carbohydrate polymer in lignocellulosic biomass is hemicellulose, which represents a family of branched carbohydrate polymers containing both pentoses (e. g. xylose, arabinose) and hexoses (e. g. galactose, glucose, mannose). Uronic acids (e. g. glucuronic acid) and **acetyl moieties** are often present as sidechain groups. The chemical composition of hemicellulose can strongly vary, depending on the botanical origin of the biomass. The degree of polymerisation is generally lower, in the range of 50 ~ 300 units. Unlike cellulose, **hemicellulose** is an **amorphous** biopolymer which is therefore more easily solubilised, and is more prone to chemical attack. The total carbohydrate fraction is often referred to as holocellulose, and includes cellulose, hemicellulose, as well as other (minor) carbohydrate bio-polymers such as pectins. Pectins account for only a small fraction of the carbohydrates in grasses, but contribute significantly to the biomass recalcitrance. Another component is lignin, the third and most complex constituent of lignocellulosic biomass. Lignin is defined as an irregular, oxygenated *p*-propylphenol polymer, formed by free radical polymerisation of monolignols in the plant cell wall, and therefore has been referred to as **supramolecular** self-assembled chaos. It provides rigidity to the plant cell wall as well as resistance to microbial attack, and thus contributes to biomass recalcitrance. Lignin biosynthesis is initiated by the phenylpropanoid pathway, a **multienzyme** biochemical network wherein phenylalanine (and tyrosine, in grasses) is converted into the main lignin building blocks: *p*-coumaryl alcohol, **coniferyl alcohol**, and **sinapyl alcohol** (Figure 7.1). These building blocks, or monolignols, differ in the amount of methoxy groups on the phenolic nucleus, which are commonly abbreviated as H (p-hydroxyphenyl), G (guaiacyl), and S (syringyl). The relative distribution of phenolic nuclei in lignin strongly differs between plant species. In general, softwood lignin (e.g. pine, spruce) exclusively contains G units, whereas hardwood lignin (e.g. birch, poplar, eucalyptus) is composed of both G and S units.

Selected from: *Chem. Soc. Rev.*, 2018, 47: 852-908.

Words and Expressions:

lignocellulose [ˌlɪgnə(ʊ)ˈseljʊləʊz; -s] *n.* [植] 木质纤维素
abundant [əˈbʌndənt] *adj.* 丰富的；充裕的；盛产
herbaceous [hɜːˈbeɪʃəs] *adj.* 草本的；绿色的；叶状的
glucose [ˈgluːkəʊs; -z] *n.* 葡萄糖；葡糖
van der Waals force 范德华力，范德华引力
acetyl moiety [ˈæsɪtaɪl; -tɪl][ˈmɒɪɪtɪ] 乙酰基
hemicellulose [hemɪˈseljʊləʊz; -s] *n.* [林] 半纤维素
amorphous [əˈmɔːfəs] *adj.* 无定形的；无组织的；[物] 非晶形的
supramolecular [ˌs(j)uːprəməˈlekjʊlə] *adj.* 超分子的(由许多分子组成的)
multienzyme [ˌmʌltiˈenzaim] *adj.* 多酶的
coniferyl alcohol 松柏醇
sinapyl alcohol 芥子醇

Notes:

1) The chemical composition of hemicellulose can strongly vary, depending on the botanical origin of the biomass.

根据生物质的植物学来源不同，半纤维素的化学组成成分可能有很大的差异。

2) Lignin is defined as an irregular, oxygenated *p*-propylphenol polymer, formed by free radical polymerisation of monolignols in the plant cell wall, and therefore has been referred to as supramolecular self-assembled chaos.

木质素是一种不规则的、含氧的对丙苯酚基聚合物。这类聚合物是由植物细胞壁中木质素单体通过自由基聚合而成。因此，木质素也被称为是无序组装的超分子结构。

Recommended Reading Materials:

1. Terashima N, Kitano K, Kojima M, Yoshida M, Yamamoto H, Westermark U. Nanostructural assembly of cellulose, hemicellulose, and lignin in the middle layer of secondary wall of ginkgo tracheid. *J. Wood Sci.*, 2009, 55: 409-416.

2. Aro T, Fatehi P. Production and application of lignosulfonates and sulfonated lignin. *ChemSusChem*, 2017, 10: 1861-1877.

3. Farhat W, Venditti R A, Hubbe M, Taha M, Becquart F, Ayoub A. A review of water-resistant hemicellulose-based materials: processing and applications. *ChemSusChem*, 2017, 10: 305-323.

Lesson 9 Lignocellulose Fractionation

Biomass **fractionation** technology is located at the heart of the **biorefinery** (Figure 9.1). It determines the fate of the individual lignocellulose constituents, and therefore, it can both widen or limit the array of downstream valorisation possibilities, especially regarding lignin (i.e. because of structural alteration, **sulfur** incorporation, etc.). Many fractionation **approaches** have been developed to date, ranging from traditional paper making to more sophisticated and environmentally friendly innovations. Each of these methods results in a specific lignin product, which can be isolated in the form of (i) a solid residue, (ii) a lignin precipitate, or even directly as (iii) a **depolymerised** product mixture.

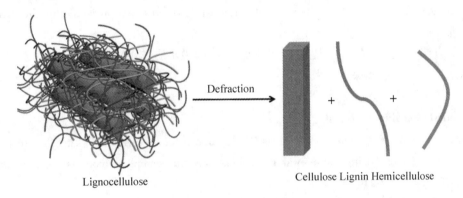

Figure 9.1 Schematic illustration of defraction of lignocellulosic biomass

The following section discusses the extensive and diverse domain of biomass fractionation, with particular attention to the associated lignin characteristics. The various fractionation technologies are divided into two distinct classes. The first class covers methods that focus on the liberation of lignin from the biomass matrix (i.e. delignification), while the (hemi)cellulose carbohydrates are preserved in the form of a delignified **pulp**. Depending on the particular method, the lignin is **isolated** as a solid lignin **precipitate** (LP) or as a depolymerised lignin oil (DL). The second class of lignocellulose fractionation **strategies** comprises methods that target the conversion and solubilisation of the carbohydrate fractions. Herein, lignin is mostly isolated in the form of an insoluble lignin residue (LR) or as a lignin precipitate (LP).

Selected from: *Chem. Soc. Rev.*, 2018, 47: 852-908.

Words and Expressions:

fractionation [ˌfrækʃənˈeʃən] n. 分别；分馏法

biorefinery 生物炼制

sulfur ['sʌlfɚ] vt. 用硫黄处理；n. 硫黄；硫黄色
approach [ə'prəʊtʃ] n. 方法；途径；接近
sophisticated [sə'fɪstɪkeɪtɪd] adj. 复杂的；精致的
depolymerise [diː'pɒlɪməraɪz] vt. 使解聚；vi. 解聚
pulp [pʌlp] n. 纸浆；果肉；黏浆状物质
isolated ['aɪsəleɪtɪd] adj. 孤立的；分离的；单独的；[电] 绝缘的
precipitate [prɪ'sɪpɪteɪt] n. [化学] 沉淀物
strategy ['strætədʒɪ] n. 战略，策略

Notes：

1）Many fractionation approaches have been developed to date, ranging from traditional paper making to more sophisticated and environmentally friendly innovations.

到目前为止，（人们）已经开发了许多分离方法。这些方法包括传统造纸中的分离方法、也包括更复杂和对环境友好的新型分离方法。

2）The first class covers methods that focus on the liberation of lignin from the biomass matrix (i.e. delignification), while the (hemi) cellulose carbohydrates are preserved in the form of a delignified pulp.

第一类方法侧重于从生物质基质中分离木质素（即脱木质作用），而（半）纤维素类碳水化合物则以脱木素浆的形式保存下来。

Recommended Reading Materials：

1. Lancefield C S, Panovic I, Deuss P J, Barta K, Westwood N J. Pre-treatment of lignocellulosic feedstocks using biorenewable alcohols: towards complete biomass valorisation. *Green Chem.*, 2017, 19: 202-214.

2. Rinaldi R, Jastrzebski R, Clough M T, Ralph J, Kennema M, Bruijnincx P C, Weckhuysen B M. Paving the way for lignin valorisation: recent advances in bioengineering, biorefining and catalysis. *Angew. Chem. Int. Ed.*, 2016, 55: 8164-8215.

Lesson 10 Cellulose: Structure and Properties

Cellulose is the major component in lignocellulosic plant biomass and is approximately 40% ~ 45% of wood by weight, depending on wood species. It is the most abundant natural polymer on earth. Approximately 75 ~ 100 billion tons of cellulose can be produced annually worldwide. Cellulose was first isolated by Anselme Payen in 1938 when he treated wood with nitric acid. Cellulose consists of **chains** of thousands of **D-glucan** units connected by the β-1-4 linkage (Figure 10.1). An elementary cellulose fibril in plant biomass consists of multiple cellulose chains.

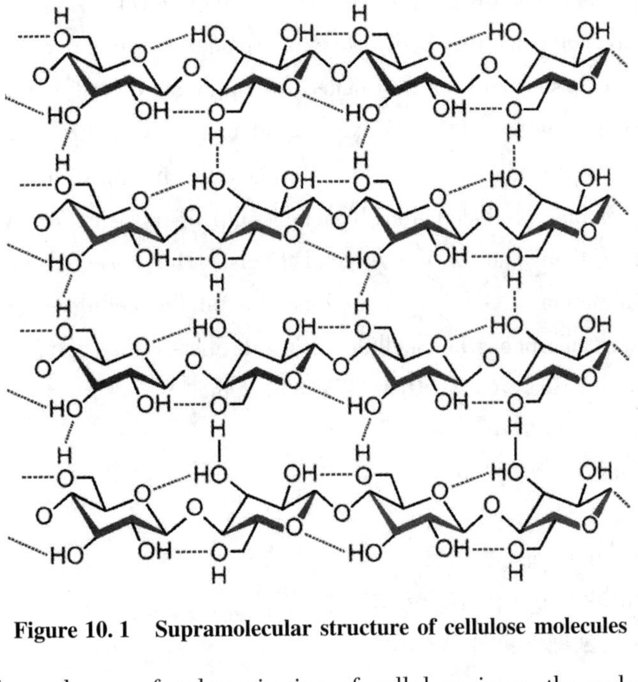

Figure 10.1 Supramolecular structure of cellulose molecules

The chain length or degree of polymerization of cellulose is on the order of 10 000 in wood cellulose. Cellulose chains are well organized and **interconnected** by hydrogen bonding, i.e., from $O(6)$ to $O(2)H$ and from $O(3)H$ to the ring $O(5)$. **Hydrogen bonding** also exists between cellulose chains, such as from $O(3)$ to $O(6)H$. These internal hydrogen bonds provide cellulose with a stable structure and impart poor solubility in water as well as many common solvents. The degree of hydrogen bonding and the local conformation of the $C(6)H_2OH$ group vary with the state of cellulose. In wood, the natural state of cellulose, called cellulose I, is less strongly hydrogen bonded and may not be locally highly ordered. In contrast, **regenerated cellulose** (cellulose II) derived from chemical dissolution and redeposition or in a process called mercerization exhibits a modified hydrogen-bonding pattern and is usually more stable compared with cellulose I. The proportion of ordered and disordered cellulose depends on the origin of the samples and the chemical processes used to prepare them. The term, cellulose allomorph, refers to cellulose materials with different crystalline structures. Although cellulose was discovered more than 175 years ago, the detailed cellulose supermolecular or ultramolecular structure is still a subject of debate. The penetration depth of water through the various structural levels of cellulose fibrils is still an open question without definitive answers. Different models describing cellulose organization within a microfibril have been proposed. Of the two commonly accepted models, one is a **disordered** (amorphous, or para-crystalline, or water accessible) region of cellulose that exists between two regions of highly ordered (crystalline or water inaccessible) cellulose. The structural size of the region described in this model is on the order of elementary fibrils, i.e., water is only accessible to the surface of elementary fibrils. Using controlled acid hydrolysis, one can isolate crystallites from natural cellulose (or cellulose I) with **dimensions** speculated to be reflective of crystalline domains within elementary fibrils by selectively **hydrolyzing** the water-accessible crystal defects and disordered regions. However, the concept of isolating pre-existing crystallites in natural cellulose microfibrils was recently challenged by the hypothesis that water penetrates wood at the cellulose chain level, and therefore, the elemental crystallite structure may not exist in wood. This new model is supported by the observation that cellulose nanocrystals (CNCs) could not be isolated from untreated wood and of water accessibility in wood cellulose samples as measured by D_2O exchange detected by Raman scattering at 1380 cm^{-1}. The **Raman band** at 1380 cm^{-1} is due to $C(6)H_2OH$ **bending** modes. Furthermore, the crystallinities of CNC samples produced from bleached pulp fibers using strong acid hydrolysis were not substantially greater than that of the original fibers. The observed increase in crystallinity measured by X-ray diffraction (XRD) may well be due to the cellulose enrichment in the CNC samples after hydrolyzing amorphous hemicelluloses in the fibers.

Selected from: *Chem. Soc. Rev.*, 2018, 47: 852-908.

Words and Expressions:

chain [tʃeɪn] *n.* 链

D-glucan [ˈgluːkən] *n.* 葡聚糖

interconnect [ɪntəkəˈnekt] *vi.* 互相联系

hydrogen bonding 氢键；氢键结合

regenerated cellulose 再生纤维素

disordered [dɪsˈɔːdəd] *adj.* 混乱的；失调的

dimensions [dɪˈmenʃənz] *n.* 规模，大小

hydrolyze [ˈhaɪdrəlaɪz] *v.* 水解

Raman band 拉曼带

bending [ˈbendɪŋ] *n.* 弯曲度；*v.* 弯曲

Notes:

1) In wood, the natural state of cellulose, called cellulose I, is less strongly hydrogen bonded and may not be locally highly ordered.

在木材中，纤维素的自然状态，称为纤维素 I，氢键作用较弱。因此，它们很可能在局部不是高度有序的排列。

2) The term, cellulose allomorph, refers to cellulose materials with different crystalline structures. Although cellulose was discovered more than 175 years ago, the detailed cellulose supermolecular or ultramolecular structure is still a subject of debate.

纤维素结晶变体是指具有不同晶体结构的纤维素材料。尽管纤维素是175多年前被发现的，但其具体的超分子态或超分子结构仍是一个有争议的话题。

3) The structural size of the region described in this model is on the order of elementary fibrils, i. e., water is only accessible to the surface of elementary fibrils.

在该模型中，所描述的区域结构尺寸为原细纤维级的。这意味着水只能到达其原细纤维的表面。

Recommended Reading Materials:

1. Abe K, Yano H. Comparison of the characteristics of cellulose microfibril aggregates isolated from fiber and parenchyma cells of Moso bamboo (*Phyllostachys pubescens*). *Cellulose*, 2010, 17: 271-277.

2. Abe K, Yano H. Comparison of the characteristics of cellulose microfibril aggregates of wood, rice straw and potato tuber. *Cellulose*, 2009, 16: 1017-1023.

Lesson 11 Solution of Cellulose in Ionic Liquid

Cellulose is the most abundant polymer in lignocellulose (35%-50%). Understanding its interactions with ionic liquids is thus important for the processing of lignocellulosic biomass in ionic liquids; therefore we will include a brief survey of this area here. Dissolution of cellulose is desired in biomass-to-fuels processing as well as for the production of man-made cellulose fibres. Only a few solvents are able to dissolve the crystalline polymer fibrils, e. g. N-methylmorpholine-N-oxide (NMO) or concentrated phosphoric acid, none of which are applied in lignocellulose processing. The Viscose and Lyocell processes are used for the commercial processing of pure cellulose (e. g. cotton). In the Viscose process, the cellulose is solubilised chemically by transforming the hydroxyl groups into xanthate esters using **carbon disulfide** (CS_2), which makes the cellulose soluble in organic solvents. The cellulose is regenerated by hydrolysing these esters. This process uses toxic and highly **flammable** CS_2 and produces large amounts of waste products, which has fuelled the search for a more benign replacement. The Lyocell process is a true dissolution process using NMO and an improvement in terms of health and environmental impact. **Nevertheless**, significant instability of the NMO/cellulose solution above 90 ℃ and redox activity of the NMO have been viewed as problems, which could be addressed by the use of more stable cellulose dissolving ionic liquids.

It has been known for some time that 1-ethylpyridiniumchloride, [C_2 pyr] Cl, can dissolve cellulose. More recently, it has been demonstrated that a range of ionic liquids, typically with 1,3-dialkylimidazolium cations, are also effective cellulose solvents. Clear viscous solutions are obtained, showing the typical behaviour of polymer solutions in general and solutions of cellulose in particular. This has sparked great interest. In the past few years, a considerable number of studies have been published on cellulose solubility in ionic liquids, patents have been granted for cellulose solubilisation using ionic liquids, and the literature has been reviewed several times. Many of these reviews contain **comprehensive** tables listing the **substantial** number of solubilisation studies for cellulose (and often other biopolymers), we therefore direct the reader to use these studies should they be interested in such information.

Cellulose dissolution is an industrially attractive application of ionic liquids, due to good solubilities (5%-20 wt%, depending on the ionic liquid and the conditions), the competitive properties of cellulose reprecipitated from the ionic liquid solutions, the increased stability of ionic liquid cellulose solutions and the low toxicity of certain relevant ionic liquids (for example, [$C_2 C_1 im$][$MeCO_2$] is classified as non-toxic and non-irritant). The dissolved cellulose can be modified in solution or **regenerated** (reprecipitated) by adding water, mixtures of water with organic solvents (e. g. acetone) or protic organic solvents, such as ethanol to form films and fibres. The ordering of the regenerated cellulose is reduced compared to the initial state and it is transformed into cellulose II. This also results in significantly **accelerated** hydrolysis with cellulases compared to native cellulose, an effect that is very attractive in terms of the biorefinery and has sparked interest in the use of cellulose dissolving ionic liquids in lignocellulose deconstruction.

Selected from: *Green Chemistry*, 2013, 15: 550-583.

Words and Expressions:
  carbon disulfide　[ˈkɑːrbən][daɪˈsʌlfaɪd]　[无化]二硫化碳
  flammable　[ˈflæməbl]　*adj.* 易燃的；可燃的；可燃性的
  nevertheless　[ˌnevərðəˈles]　*adv.* 然而，不过；虽然如此
  sparked　[spɑːrkt]　*v.* 点燃，发动(spark 的过去式，过去分词)
  comprehensive　[ˌkɑːmprɪˈhensɪv]　*adj.* 综合的；广泛的；有理解力的
  substantial　[səbˈstænʃl]　*adj.* 大量的；实质的；内容充实的
  regenerated　[rɪˈdʒenərɪt]　*adj.* 再生的
  accelerated　[əkˈseləreɪtɪd]　*adj.* 加速的；加快的

Notes:
1) The Viscose and Lyocell processes are used for the commercial processing of pure cellulose (e. g. cotton).
黏胶和细胞溶解工艺用于纯纤维素(例如棉)的商业加工。

2) The dissolved cellulose can be modified in solution or regenerated (reprecipitated) by adding water, mixtures of water with organic solvents (e. g. acetone) or protic organic solvents, such as ethanol to form films and fibres.
溶解的纤维素在溶剂中可以被修饰或者加水再生。通过引入水或者对应的有机溶剂，例如乙醇，可将溶解的纤维素变成膜或者纤维丝材料。

Recommended Reading Materials:
1. Zhang J, Wu J, Yu J, Zhang X, He J, Zhang J. Application of ionic liquids for dissolving cellulose and fabricating cellulose-based materials: state of the art and future trends. Materials Chemistry Frontiers, 2017, 1.7: 1273-1290.

2. Ren Q, Wu J, Zhang J, He J, Guo M. Synthesis of 1-allyl, 3-methylimidazolium-based room-temperature ionic liquid and preliminary study of its dissolving cellulose. Acta Polymerica Sinica, 2003, 3: 448-451.

Lesson 12　Cellulose: Chemical Modification

Unmodified cellulose has a low **performance** as well as variable **physical stability**. However, a chemical modification of cellulose can be executed to achieve **adequate** structural **durability**. The properties of cellulose, such as its **hydrophilic** or **hydrophobic** character, **elasticity**, water **sorbency**, adsorptive or ion exchange capability, resistance to microbiological attack and thermal resistance, are usually modified by chemical treatments. The β-D glucopyranose on the cellulose chain contains one primary **hydroxyl** group and two secondary hydroxyl groups. Functional groups may be attached to these hydroxyl groups through a variety of reactions (Figure 12.1). The main routes of direct cellulose modification in the preparation of adsorbent materials are **esterification**, **etherification**, **halogenations**, oxidation and **alkali** treatment.

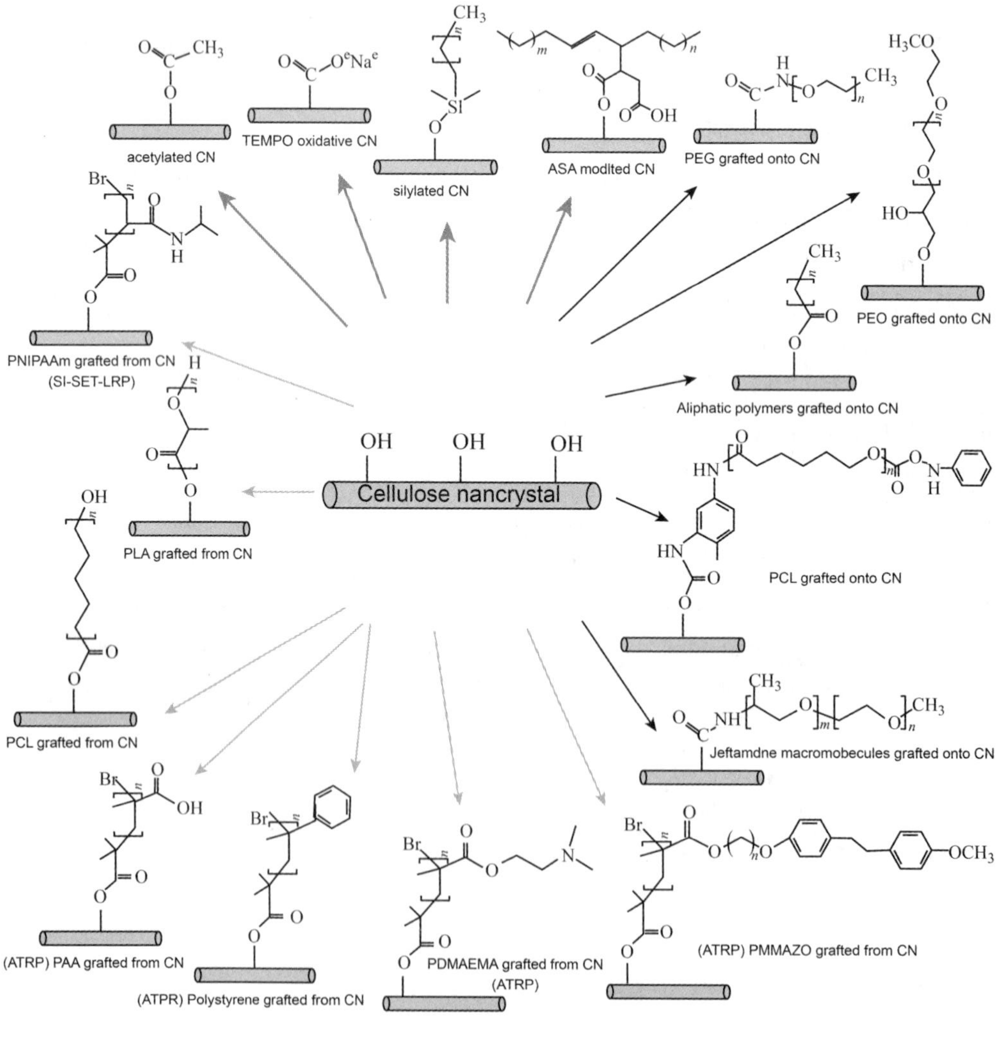

Figure 12.1　Chemical modification of cellulose

Esterification. Cellulose esters are cellulose derivatives which result from the esterification of free hydroxyl groups of the cellulose with one or more acids, whereby cellulose reacts as a trivalent polymeric alcohol. Cellulose esters are commonly derived from natural cellulose by reacting with organic acids, anhydrides, or acid chlorides. The esterification methods of cellulose leading to adsorbent materials for water treatment. The treatments of cellulose with cyclic anhydrides, such as succinic anhydride, are widely studied methods to add carboxyl groups to the surface of cellulose. EDTA dianhydride, citric acid anhydride and maleic anhydride were also used for esterification.

Silynation. Silane-based surface modification is a popular way to modify cellulosic fibers. Therefore, interactions of silane coupling agents with natural fibers are well-known. Coupling agents usually improve the degree of crosslinking in the interface region and offer a perfect bonding. Silanes undergo hydrolysis, condensation, and a bond formation stage. Silanols can form polysiloxane structures by reacting with a hydroxyl group of the cellulose fibers. In the presence of moisture, the hydrolyzable alkoxy group leads to the formation of silanols. The silanols then react with the hydroxyl groups of the fiber, forming stable covalent bonds to the cell wall.

Selected from *Water Res*., 2016, 91: 156-173.

Words and Expressions:

  unmodified [ʌnˈmɒdɪfaɪd] *adj*. 未更改的，未变的
  performance [pəˈfɔːm(ə)ns] *n*. 性能；绩效
  physical stability 物理稳定性
  adequate [ˈædɪkwət] *adj*. 充足的；适当的；胜任的
  durability [ˌdjʊərəˈbɪləti] *n*. 耐久性；坚固；耐用年限
  hydrophilic [ˌhaɪdrə(ʊ)ˈfɪlɪk] *adj*. [化学] 亲水的
  hydrophobic [haɪdrə(ʊ)ˈfəʊbɪk] *adj*. 疏水的
  elasticity [elæˈstɪsɪtɪ] *n*. 弹性；弹力；灵活性
  hydroxyl [haɪˈdrɒksaɪl; -sɪl] *n*. 羟基，氢氧基
  esterification [ɛˌstɛrəfəˈkeʃən] *n*. [有化] 酯化 (作用)
  etherification [iˌθe-rifiˈkeɪʃən] *n*. [有化] 醚化；醚作用
  halogenations [ˌhælədʒiˈneɪʃən] *n*. 卤化，加卤作用
  alkali [ˈælkəlaɪ] *n*. 碱；可溶性无机盐；*adj*. 碱性的

Notes:

1) Cellulose esters are cellulose derivatives which result from the esterification of free hydroxyl groups of the cellulose with one or more acids, whereby cellulose reacts as a trivalent polymeric alcohol.

纤维素酯是纤维素的衍生物，它是由纤维素的游离羟基与一种或多种酸酯化而成。在反应的过程中，纤维素作为三价聚合醇参与反应。

2) Silanes undergo hydrolysis, condensation, and a bond formation stage. Silanols can form polysiloxane structures by reacting with a hydroxyl group of the cellulose fibers.

硅烷经历水解、缩合和成键阶段。硅烷醇可与纤维素纤维的羟基反应形成聚硅氧烷结构。

Recommended Reading Materials:

1. O'Connell D W, Birkinshaw C, O'Dwyer T F. Heavy metal adsorbents prepared from the modification of cellulose: a review. *Bioresour. Technol.*, 2008, 99: 6709-6724.

2. Saito T, Isogai A. Ion-exchange behavior of carboxylate groups in fibrous cellulose oxidized by the TEMPO-mediated system. *Carbohydr. Polym.*, 2005, 61: 183-190.

Lesson 13　Cellulose Nanocrystals

A controlled strong acid **hydrolysis** treatment can be applied to cellulosic fibers allowing dissolution of amorphous domains and therefore **longitudinal** cutting of the microfibrils. The ensuing nanoparticles are generally called cellulose nanocrystals (CNCs) and are obtained as an **aqueous** suspension. When observed between crossed-Nicols the CNC dispersion shows the formation of **birefringent** domains. During the acid hydrolysis process, the hydronium ions penetrate the cellulose chains in the amorphous regions promoting the hydrolytic cleavage of the glycosidic bonds and releasing individual crystallites after mechanical treatment (sonication) (Figure 13.1). Different strong acids have been shown to successfully **degrade** cellulose fibers, but **hydrochloric** and **sulfuric** acids have been extensively used. However, **phosphoric**, **hydrobromic** and **nitric acids** have also been reported for the preparation of crystalline cellulosic nanoparticles. One of the main reasons for using sulfuric acid as hydrolyzing agent is its reaction with the surface hydroxyl groups via an esterification process allowing the grafting of **anionic** sulfate ester groups. The presence of these negatively charged groups induces the formation of a negative electrostatic layer covering the nanocrystals and promotes their dispersion in water.

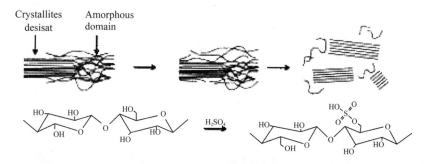

Figure 13.1　**Hydrolysis of cellulose**

However, it compromises the **thermostability** of the nanoparticles. To increase the thermal stability of H_2SO_4-prepared nanocrystals, **neutralization** of the nanoparticles by sodium hydroxide (NaOH) can be carried out. These nanoparticles occur as high aspect ratio rod-like nanocrystals, or whiskers. Their geometrical dimensions depend on the origin of the cellulose **substrate** and hydrolysis conditions. CNCs obtained from two different cellulosic sources. Each rod can be considered as a cellulose crystal with no apparent defect. CNCs generally present a relatively broad distribution in length because of the diffusion-controlled nature of the acid hydrolysis. The average length is generally of the order of a few hundreds **nanometers** and the width is of the order of a few nanometers. An important parameter for CNCs is the aspect ratio, which is defined as the ratio of the length to the width. It varies between 10 for cotton and 67 for tunicin or capim dourado (golden grass). Acid hydrolysis is the classical way of preparing CNCs. However, other processes allowing the release of crystalline domains from cellulosic fibers have more recently been reported, including enzymatic hydrolysis treatment, TEMPO oxidation, hydrolysis with gaseous acid, and treatment with ionic liquids.

Selected from: *Mater. Today*, 2013, 16: 220-227.

Words and Expressions:

  hydrolysis [ˌhaɪˈdrɒlɪsɪs] *n.* 水解作用
  longitudinal [ˌlɒn(d)ʒɪˈtjuːdɪn(ə)l] *adj.* 长度的，纵向的；经线的
  aqueous [ˈeɪkwɪəs] *adj.* 水的，水般的
  birefringent [ˌbaɪriˈfrɪndʒənt] *adj.* [光] 双折射的
  degrade [dɪˈɡreɪd] *v.* 降级，降低；退化；降解
  hydrochloric [ˌhaɪdrəʊˈklɒrɪk] *adj.* 氯化氢的，盐酸的
  sulfuric [sʌlˈfjʊərɪk] *adj.* 硫黄的；含多量硫黄的；含(六价)硫的
  phosphoric [fɒsˈfɒrɪk] *adj.* 磷的，含磷的
  hydrobromic [ˌhaɪdrəʊˈbrəumɪk] *adj.* 氢溴酸的；溴化氢的
  nitric acids [ˈnaɪtrɪk] 硝酸
  anionic [ˌænaɪˈɒnɪk] *adj.* 阴离子的，带负电荷的离子的
  thermostability [θɜːmostəˈbɪləti] *n.* [热] 热稳定性；耐热性
  neutralization [ˌnjuːtrəlaɪˈzeɪʃən] *n.* [化学] 中和；[化学] 中和作用
  substrate [ˈsʌbstreɪt] *n.* 基质；基片；底层
  nanometer [ˈneɪnəˌmiːtə] *n.* [计量] 纳米(即十亿分之一米)

Notes:

1) One of the main reasons for using sulfuric acid as hydrolyzing agent is its reaction with the surface hydroxyl groups via an esterification process allowing the grafting of anionic sulfate ester groups.

采用硫酸作为水解剂的主要原因之一是其可以（在水解的同时）与表面羟基发生磺化反应，在表面引入磺酸酯基团。

2) However, other processes allowing the release of crystalline domains from cellulosic fibers have more recently been reported, including enzymatic hydrolysis treatment, TEMPO oxidation, hydrolysis with gaseous acid, and treatment with ionic liquids.

然而，其他报道的从纤维素中制备具有晶体结构域的纤维的方法有酶水解、氧化、气态酸水解与离子液体处理。

Recommended Reading Materials:

1. Elazzouzi-Hafraoui S, Nishiyama Y, Putaux J L, Heux L, Dubreuil F, Rochas C. The shape and size distribution of crystalline nanoparticles prepared by acid hydrolysis of native cellulose. Biomacromolecules, 2008, 9: 57-65.

2. Filpponen I, Argyropoulos D S. Regular linking of cellulose nanocrystals via click chemistry: synthesis and formation of cellulose nanoplatelet gels. Biomacromolecules, 2010, 11: 1060-1066.

Lesson 14　　Cellulose Hydrogel

Gels are defined as three-dimensional polymer networks swollen by large amounts of solvent. **Hydrogels** are, mainly, structures formed from biopolymers and/or **polyelectrolytes**, and contain large amounts of **trapped** water. Concerning definitions of hydrogel types, according to the source, hydrogels can be divided into those formed from natural polymers and those formed from synthetic polymers. On the basis of the cross-linking method, the hydrogels can be divided into chemical gels and physical gels.

Physical gels are formed by molecular self-assembly through ionic or hydrogen bonds, while chemical gels are formed by covalent bonds. Hydrogels were first reported by Wichterle and Lím. It is worth noting that the hydrogels have wide potential applications in the fields of food, biomaterials, **agriculture**, **water purification**, etc. Recently, scientists have devoted much energy to developing novel hydrogels for applications such as **biodegradable** materials for drug delivery, tissue engineering, sensors, **contact lenses**, purification, etc. Synthetic polymer-based hydrogels have been reported such those formed by cross-linking **poly(ethylene glycol)**, **poly(vinyl alcohol)**, poly(amido-amine), poly(N-isopropylacrylamide), polyacrylamide, and poly(acrylic acid) and their copolymers. Synthetic hydrogels like PEG-based hydrogels have advantages over natural hydrogels, such as the ability for **photopolymerization**, adjustable mechanical properties, and easy control of scaffold architecture and chemical compositions, but PEG hydrogels alone cannot provide an ideal environment to support cell adhesion and tissue formation due to their bio-inert nature.

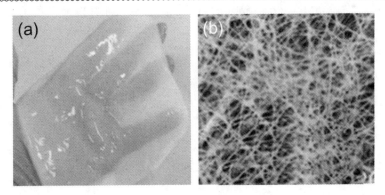

Figure 14.1　(a) Images of cellulose hydrogel and　(b) TEM images of cellulose hydrogel

A number of **polysaccharides** have similar properties to PEG in terms of biocompatibility and low protein and cell adhesion, and they can been biodegraded to nontoxic products that are easily assimilated by the body (Figure 14.1). Various hydrogels from natural polymers have been fabricated by using hyaluronate, alginate, starch, **gelatin**, cellulose, chitosan, and their derivatives, showing potential application in biomaterials field because of their safety, hydrophilicity, **biocompatibility** and biodegradability. Cellulose, the most abundant renewable resource on earth, will become the main chemical resource in the future. Moreover, numerous new functional materials from cellulose are being developed over a broad range of applications, because of the increasing demand for environmentally friendly and biocompatible products. Cellulose having abundant hydroxyl groups can be used to prepare hydrogels easily with fascinating structures and properties. There is a need to study cellulose-based hydrogels in both fundamental research and industrial application. Cellulose hydrogels can be prepared from a cellulose solution through physical cross-linking. Because cellulose has many hydroxyl groups which can form hydrogen bonding linked **network** easily. However, cellulose is very difficult to be dissolved in common solvents due to its highly extended hydrogen bonded structure, so the major problem for preparing cellulose hydrogel is a lack of appropriate solvents. Recently, new solvents, such as N-methylmorpholine-N-oxide (NMMO), ionic liquids (ILs), and alkali/urea (or thiourea) aqueous systems have been developed to dissolve cellulose, providing great opportunities for the preparation of cellulose hydrogels. Bacterial cellulose (BC) is also a strong candidate for the **fabrication** of cellulose-based hydrogels, since certain bacterial species possesses the ability to create pure cellulose hydrogel.

Selected from: *Carbohyd. Polym.*, 2011, 84: 40-53.

Words and Expressions:

gels [dʒelz] n. 凝胶剂
hydrogel [ˈhaɪdrə(ʊ)dʒel] n. [物化] 水凝胶
polyelectrolyte [ˌpɒlɪˈlektrəlaɪt] n. [物化] 聚合电解质
trapped [træpt] adj. 捕获的；收集的；受到限制的
agriculture [ˈæɡrɪkʌltʃə] n. 农业；农耕；农业生产；农艺，农学
water purification [ˌpjʊərɪfɪˈkeɪʃən] 水净化
biodegradable [ˌbaɪə(ʊ)dɪˈɡreɪdəb(ə)l] adj. 生物所能分解的，能进行生物降解的
contact lenses 隐形眼镜
poly(ethylene glycol) 聚(乙二醇)
poly(vinyl alcohol) 聚(乙烯醇)
photopolymerization [ˈfəʊtəʊˌpɒlɪməraɪˈzeɪʃən] n. 光聚合，光致聚合作用
polysaccharide [ˌpɒlɪˈsækəraɪd] n. [有化] 多糖；多聚糖
gelatin [ˈdʒelətɪn] n. 明胶；动物胶；胶制品
biocompatibility [ˌbaɪəʊkəmˌpætəˈbɪlɪti] n. 生物相容性；生物适合性
network [ˈnetwɜːk] n. 网络
fabrication [ˌfæbrɪˈkeɪʃ(ə)n] n. 制造，建造；装配；伪造物

Notes:

1) Synthetic hydrogels like PEG-based hydrogels have advantages over natural hydrogels, such as the ability for photopolymerization, adjustable mechanical properties, and easy control of scaffold architecture and chemical compositions, but PEG hydrogels alone cannot provide an ideal environment

to support cell adhesion and tissue formation due to their bio-inert nature.

PEG 等合成水凝胶相比天然水凝胶具有可通过光聚合来制备、力学性能可调、多级支架结构与化学组分可控等优点。但由于其内在的生物惰性，PEG 类的合成水凝胶无法给细胞黏附与组织形成提供理想的环境。

2) However, cellulose is very difficult to be dissolved in common solvents due to its highly extended hydrogen bonded structure, so the major problem for preparing cellulose hydrogel is a lack of appropriate solvents.

然而，纤维素由于其高度延伸的氢键结构，很难溶于普通溶剂，因此制备纤维素水凝胶的主要问题是缺乏合适的溶剂。

Recommended Reading Materials:

1. Kelly J A, Shukaliak A M, Cheung C C, Shopsowitz K E, Hamad W Y, MacLachlan M J. Responsive photonic hydrogels based on nanocrystalline cellulose. *Angew. Chem. Int. Ed.*, 2013, 52: 8912-8916.

2. Marc G, Mele G, Palmisano L, Pulito P, Sannino A. Environmentally sustainable production of cellulose-based superabsorbent hydrogels. *Green Chem.*, 2006, 8: 439-444.

3. Chang C, He M, Zhou J, Zhang L. Swelling behaviors of pH- and salt-responsive cellulose-based hydrogels. *Macromolecules*, 2011, 44: 1642-1648.

Lesson 15 Cellulose-based Aerogel Absorbers

As a "young" third generation of **aerogel** materials succeeding **silica** and synthetic polymer-based ones, cellulose aerogels or **sponges** combine the **intriguing features** of aerogel-type materials with additional advantages of naturally occurring cellulose, such as abundant sources, natural **renewability**, biodegradability, and ease to surface modification. Therefore, cellulose aerogels seem to be one of the most fascinating natural oil sorbents after appropriate modifications. Depending on the nature of cellulosic materials, cellulose aerogels include cellulose derivative-based ones, regenerated cellulose (RC)-based ones, and nanocellulose-based ones. **Prior to** the formation of cellulose aerogels by a drying process, the gelation is a key step. During the gelation, the three dimensional cellulose network (3D) is formed (Figure 15.1). Depending on whether the reaction is involved or not during the formation of gels, the gelation mechanism can be classified into the physical **cross-linking** and chemical cross-linking. For the former mechanism, the **intramolecular** and/or **intermolecular** hydrogen bonds and physical entanglement between cellulose molecules are mainly responsible for the gelation. The physical crosslinking mechanism is involved in the case of both RC aerogels and nanocelulose ones. For the chemical gelation mechanism, an additional cross-linking agent, such as paper-strengthening **resin** needs to be added to induce the formation of the crosslinked cellulose network.

Figure 15.1 (a) Images of aerogels (N, E, F and U). (b) Images of SA-based aerogels in atmospheres with different humidity, scale bar = 2 cm.

The initially developed cellulose aerogel-type oil sorbents are based on regenerated cellulose (RC). Generally, RC-basedaerogels are prepared through the following steps: (i) fully dissolving the cellulose in an appropriate solvent; (ii) regeneration of cellulose by replacing the solvent by a nonsolvent (i.e., gelation); (iii) drying of the obtained hydrogels. Nevertheless, the dissolution, gelation, and solvent-exchange steps are very time-consuming in the preparation of RC aerogels and also the used solvents are usually very harmful. During the regeneration step, the gelation mechanism mostly belongs to physical crosslinking. Moreover, unlike nanocellulose (e.g., NFC) aerogels, the favorable cellulose-I crystalline structure is converted to cellulose-II one during the preparation of RC aerogels, yielding the aerogels with relatively inferior mechanical properties (e. g., **fragility**) and lower aspect ratio of the fibrils with respect to NFC. It is well-known that the oil-sorption performance of cellulose aerogels depends not only on the density and viscosity of oily liquid, but also largely on the capillary effect, van der Waals forces, hydrophobic interaction between the oils and absorbents, and morphological parameters of the aerogels (e. g., surface wettability, total pore volume, and pore structure). The oil density would contribute to the **saturated** absorption capacity in aerogel-type sorbents with same pore size distribution and 3D network structure. And a lower oil viscosity facilitates their penetration into the porous network of aerogels, and thus results in a higher adsorption capacity, whereas highly porous aerogels usually tend to show higher oil-sorption capacities because they provide more internal free volume for oil sorption.

The pore structure of cellulose aerogels critically depends on the choice of drying processes. In practice, freeze and supercritical dryings are the most commonly used methods for the preparation of cellulose aerogels. Because a low **surface tension** effect (i. e., capillary effect) occurs during the drying stage, the supercritical carbon dioxide ($SC-CO_2$) drying can effectively avoid the collapse of 3D porous structure and result in cellulose aerogels with a low density and a high specific surface area.

Selected from: *ACS Sustainable Chem. Eng.*, 2017, 5: 49-66.

Words and Expressions:
    aerogel [ˈeərədʒel] n. [物化] 气凝胶
    silica [ˈsɪlɪkə] n. 二氧化硅；[材] 硅土
    sponge [spʌn(d)ʒ] n. 海绵；海绵状物
    intriguing feature 有趣的功能
    renewability 可再生性
    prior to 在……之前；居先
    cross-linking n. 交联；交叉结合
    intramolecular [ˌɪntrəməˈlekjʊlə] adj. [化学] 分子内的
    intermolecular [ˌɪntəməˈlekjʊlə] adj. 分子间的；作用于分子间的，存在于分子间的
    resin [ˈrezɪn] n. 树脂；松香；vt. 涂树脂；用树脂处理
    fragility [frəˈdʒɪlɪtɪ] n. 脆弱；[力] 易碎性；虚弱
    saturated [ˈsætʃəreɪtɪd] adj. 饱和的；渗透的；深颜色的
    viscosity [vɪˈskɒsɪtɪ] n. [物] 黏性，[物] 黏度
    surface tension 表面张力

Notes:
1) Generally, RC-based aerogels are prepared through the following steps: (i) fully dissolving the cellulose in an appropriate solvent; (ii) regeneration of cellulose by replacing the solvent by a nonsolvent (i. e., gelation); (iii) drying of the obtained hydrogels.

通常，再生纤维素基的气凝胶是通过以下步骤制备的：(i)将纤维素充分溶解在适当的溶

剂中；(ii)用不溶性溶剂代替可溶性溶剂再生纤维素(如凝胶)；(iii)干燥所得凝胶。

2) A lower oil viscosity facilitates their penetration into the porous network of aerogels, and thus results in a higher adsorption capacity, whereas highly porous aerogels usually tend to show higher oil-sorption capacities because they provide more internal free volume for oil sorption.

较低的油黏度有利于它们渗透到气凝胶的多孔网络中，从而产生较高的吸附能力。与此同时，高孔气凝胶往往也表现出较高的吸油能力，因为它们为油吸附提供了更多的内部自由体积。

Recommended Reading Materials：

1. Mulyadi A, Zhang Z, Deng Y. Fluorine-free oil absorbents made from cellulose nanofibril aerogels. *ACS Appl. Mater. Interfaces*, 2016, 8：2732-2740.

2. Chen W, Li Q, Wang Y, Yi X, Zeng J, Yu H, Liu Y, Li J. Comparative study of aerogels obtained from differently prepared nanocellulose fibers. *ChemSusChem*, 2014, 7：154-161.

Lesson 16 General Introduction of Lignin

Polymeric components in plants, such as cellulose, hemicelluloses, and lignin interpenetrate with each other to form complex higher order structures in living plant **organs** in the presence of an excess amount of water in nature. The history of lignin science stretches over a period of one hundred years and many scientists had made efforts effectively to exclude the lignin from wood in order to **extract** cellulose in the pulping process. Lignin has been considered as an unwelcome **by-product** and **attempts** have been made to **cultivate** wood species having a small amount of lignin content using new biotechnology. Thus, synthetic pathways of lignin in living plant organs and also **enzymatic** or chemical degradation processes have been investigated. In "Lignins", edited by Sarkanen and Ludwig, it is stated that the word *lignin* is derived from the Latin word lignum meaning wood. The amount of lignin in plants **varies** widely. In the case of wood, the amount of lignin ranges from ca. 12% to 39%, when the amount is determined according to Klason lignin analysis which is dependent on the hydrolysis and solubilization of the carbohydrate component of the lignified material. The above-method consists of two steps. Lignified material is treated with 72% sulfuric acid (cooled prior to use at 10-15℃) at 20℃ for a certain time and then followed by dilution of the acid to 3.0%, boiling to complete hydrolysis.

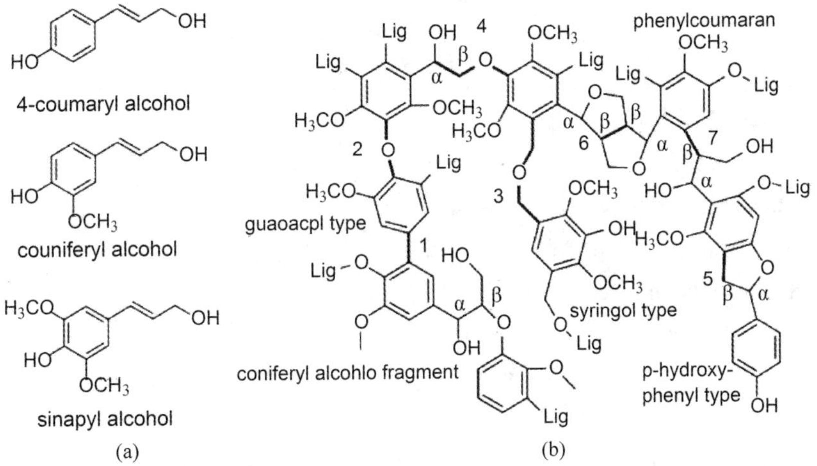

Figure 16.1 (a) monolignols and (b) lignin structure

The lignin is isolated as an acid-insoluble material. The content of acid-insoluble Klason lignin varies from ca. 29% to. 39% in soft woods and from ca. 16% to 22%. The chemical structures of lignin have been investigated in detail by chemical and **spectroscopic** methods. Lignin is usually considered as a polyphenolic material having an amorphous structure, which arises from an enzyme-initiated dehydrogenative polymerization of *p*-coumaryl, coniferyl and sinapyl alcohols. The basic lignin structure is classified into only two components; one is the aromatic part and the other is the C3 chain. The only usable reaction site in lignin is the OH group, which is the case for both phenolic and alcoholic hydroxyl groups. Lignin consists of (i) 4-hydroxyphenyl, (ii) guaiacyl, and (iii) syringyl structures connected with carbon atoms in **phenylpropanoid** units, as illustrated in Figure 16.1. It is essential to consider how to use the above basic lignin structures in synthetic pathways when lignin utilization for industrial materials is achieved. The physical chemical nature of lignin as a representative biopolymer has been neglected for a long time. Goring first shed light on the polymeric nature of lignin in early 1960. He measured the intrinsic viscosity of lignins and concluded that **conformation** of lignins is between an Einstein sphere and a non-free draining random coil in a solvent. He also showed that the **lignosulfonate** is in the form of spherical particles of a wide range of sizes according to the electron **micrograph** of high molecular weight **fraction** of sodium lignosulfonate (LS). He also reported that the softening temperature (Ts) of lignins varied from 127 to 193℃ in dry state and from 72 to 128℃ in **moist** lignins.

Selected from: *Adv. Polym. Sci.*, 2010, 232: 1-63.

Words and Expressions:

  organ [ˈɔːg(ə)n] *n.* 器官
  extract [ˈekstrækt] *vt.* 提取;取出;摘录;榨取;*n.* 汁;摘录;榨出物;选粹
  by-product [ˈbaɪ prɑdəkt] *n.* 副产品;附带产生的结果
  attempt [əˈtem(p)t] *n.* 企图,试图;攻击;*vt.* 企图,试图;尝试
  cultivate [ˈkʌltɪveɪt] *vt.* 培养;陶冶;耕作
  enzymatic [ˌenzaɪˈmætɪk] *adj.* 酶的
  vary [ˈveərɪ] *vi.* 变化;变异;*vt.* 改变;使多样化;变奏
  spectroscopic [ˌspɛktrəˈskɑpɪk] *adj.* 光谱学的;分光镜的
  phenylpropanoid 苯基丙烷类
  conformation [kɒnfɔːˈmeɪʃ(ə)n] *n.* 构造;一致,符合;构象
  lignosulfonaten 木质素磺酸盐
  micrograph [ˈmaɪkrə(ʊ)grɑːf] *n.* 显微照片,显微图
  fraction [ˈfrækʃ(ə)n] *n.* 分数;部分;小部分;稍微
  moist [mɒɪst] *adj.* 潮湿的;多雨的

Notes:

1) The amount of lignin in plants varies widely. In the case of wood, the amount of lignin ranges from ca. 12% to 39%, when the amount is determined according to Klason lignin analysis which is dependent on the hydrolysis and solubilization of the carbohydrate component of the lignified material.

植物中木质素的含量差别很大。以木材为例,木质素的含量约为12%~39%,这一数据是由 Klason 木素分析法得出。该分析法主要是对木质化原料中碳水化合物成分进行水解,增加其溶解性(然后对其中的残余固态木质素组分进行分析)。

2) Lignin is usually considered as a polyphenolic material having an amorphous structure, which

arises from an enzyme-initiated dehydrogenative polymerization of *p*-coumaryl, coniferyl and sinapyl alcohols.

木质素通常被认为是一种具有无定形结构的多酚材料,由酶引发的对香豆素、松香素和芥子醇的脱氢聚合而成。

Recommended Reading Materials:

1. Boeriu C G, Bravo D, Gosselink R J A, van Dam J E G. Characterisation of structure-dependent functional properties of lignin with infrared spectroscopy. *Ind. Crops Prod.*, 2004, 20: 205-218.

2. Kai D, Tan M J, Chee P L, Chua Y K, Yap Y L, Loh X J. Towards lignin-based functional materials in a sustainable world. *Green Chem.*, 2016, 18: 1175-1200.

Lesson 17 Characterization Techniques of Lignin (I)

Concentrations of lignin from wood pulp samples can be determined both by **non-invasive** and invasive methods. The non-invasive methods are based on the fact that chemical structure of lignin allows them to absorb **electromagnetic radiation** in specific region. The production of characteristic spectra in a specific region (wavelength, wave 15 number or chemical shift) features will be proportional to amount of lignin in sample is determined by either utilizing molar extinction coefficient (UV-visible spectroscopy), or **overlap** intensities of **modified** and unmodified matrix infrared (IR) and **near-infrared spectroscopy** (NIR) or integration of specific peaks in solid state nuclear magnetic resonance (NMR) spectra with the sample of known lignin content. The non-invasive methods dictate whether the extraction of lignin from wood is economical and cost-effective. On the other hand, invasive methods are based on volumetric titrations or gravimetric techniques using specific chemical treatments such as acetyl bromide or thioglycolate. The lignin content was estimated by the gravimetric Klason procedure. The lignin structure can be investigated by chemical methods such as thioacidolysis, **Copper oxide** oxidation, **nitrobenzene oxidation** (NBO), and derivatization followed by reductive cleavage (DFRC). The composition (H/G/S) of the lignin polymer and its quantification was achieved by DFRC method using pyrolysis-gas chromatography-mass spectrometry (GC-MS). This involved nitrobenzene oxidation, pyrolysis (GC-MS), thioacidolysis and DFRC. This analytical process involves lots of time for preparation and analysis due to multi-step process. A streamlined thioacidolysis method and nearinfrared reflectance-based prediction modeling allows quicker analysis. A spectroscopic technique such as FTIR, NMR etc. is used to differentiate the nature of wood. FTIR spectra of softwood box (*Buxus sempervirens*) and hardwood aspen (*Populus tremula*), in the fingerprint region 1800-800 cm^{-1}, showed prominent differences in the **transmittance** values. A reduced intensity of the band at 1740 cm^{-1} is slightly higher in aspen than in box which can be attributed to a greater number of **acetyl groups** in case of former. The difference in the guaiacyl content between softwood and hardwood is elaborated by a doublet detected at 1610-1595 cm^{-1}, while only one band at 1595 cm^{-1} respectively. Generally hardwood shows an equal intense peaks at 1595 and 1510 cm^{-1} attributed to the predominant syringyl unit, while in the softwood the band at 1510 cm^{-1} is more intense than at 1595 cm^{-1}, attributable to a higher content of guaiacyl units. In softwood kraft lignin (SKL), the 1269 cm^{-1} band (guaiacyl ring 55 breathing with carbonyl stretching) is more intense than the 1214 cm^{-1} band and has no syringyl absorption at 1327 cm^{-1}, whereas the opposite is true for hardwood lignins, that is, a weak 1269 cm^{-1} band, a strong band at 1215 cm^{-1}, and a syringyl absorption at around 1327 cm^{-1}. The presence of a syringyl unit in hardwood lignin is also evident from the higher intensity of the band at 1462 cm^{-1}. The presence of higher percentage of methoxy groups in hardwood is indicated by the peak near 1600 cm^{-1} due to **aromatic**-OCH$_3$ stretching.

Selected from: *RSC Adv.*, 2014, 4: 21712 – 21752.

Words and Expressions:

  non-invasive [ˌnɑːn ɪnˈveɪsɪv] *adj.* 非侵入性的；非入侵的
  electromagnetic radiation [ɪˌlektroʊmæɡˈnetɪk] [ˌreɪdiˈeɪʃn] [物] 电磁辐射
  overlap [ˌoʊvərˈlæp] *n.* 重叠的部分，互搭量
  modified [mɑdəˌfaɪd] *adj.* 改进的，修改的；改良的
  near-infrared spectroscopy [nɪr ˌɪnfrəˈred] [spekˈtrɑːskəpi] 近红外光谱学
  copper oxide [ˈkɑːpər] [ˈɑːksaɪd] [无化] 氧化铜
  nitrobenzene oxidation [ˌnaɪtrəbɛnˈzin] [ˌɑːksɪˈdeɪʃn] 硝基苯氧化
  transmittance [trænsˈmɪtəns] *n.* [光] 透射比；透明度
  acetyl groups [əˈsitəl] [ɡruːp] [有化] 乙酰基
  aromatic [ˌærəˈmætɪk] *adj.* 芳香的，芬芳的；芳香族的

Notes:

1) On the other hand, invasive methods are based on volumetric titrations or gravimetric techniques using specific chemical treatments such as acetyl bromide or thioglycolate.

另一方面，侵入性方法是使用如乙酰溴或硫代乙醇酸盐的化学物质进行（木质素）容量滴定或重量滴定。

2) In softwood kraft lignin (SKL), the 1269 cm^{-1} band is more intense than the 1214 cm^{-1} band and has no syringyl absorption at 1327 cm^{-1}, whereas the opposite is true for hardwood lignins, that is, a weak 1269 cm^{-1} band, a strong band at 1215 cm^{-1}, and a syringyl absorption at around 1327 cm^{-1}.

在软木牛皮纸木质素中，1269 cm^{-1} 吸收带比 1214 cm^{-1} 吸收带更强烈，且没有在 1327 cm^{-1} 的紫丁香基吸收（特征峰）。而同种硬木木质素与之相反，在 1269 cm^{-1} 处有弱吸收带，在 1215 cm^{-1} 有一个强的吸收带，在 1327 cm^{-1} 处呈现出紫丁香基的特征吸收峰。

Recommended Reading Materials:

1. Clifford D J, Carson D M, McKinney D E, Bortiatynski J M, Hatcher P G. A new rapid technique for the characterization of lignin in vascular plants: thermochemolysis with tetramethylammonium hydroxide (TMAH). *Organic Geochemistry*, 1995, 23(2): 169-175.

2. Hedges J I, Ertel J R. Characterization of lignin by gas capillary chromatography of cupric oxide oxidation products. *Analytical Chemistry*, 1982, 54(2): 174-178.

Lesson 18　Characterization Techniques of Lignin (II)

NMR spectroscopy provides information about the **structural configuration**, **quantification**, chemical composition, and linkages present in lignin sample. ^{13}C-^{1}H correlated (HSQC, HMQC), and ^{13}C-NMR both solution and solids state are reported in literature to elucidate the structure of lignin. NMR studies confirmed the higher concentration of methoxy signals in hardwood Kraft lignin (HKL) as compared to SKL due to **predominance** of both G and S units in the former. It is observed that purified isolated lignins, namely, "**cellulolytic enzyme lignin**" gave good quality spectra as they are devoid of cellulosic component. Cell wall polymers lignin and polysaccharides in native state are also identified and characterized during hydrothermal treatment of wheat straw lignin using solution state 2D-NMR spectroscopy. Non-woody biomass such as corn stover was studied for the structural changes showed by lignin and LCCs characterized by alkaline nitrobenzene oxidation, ^{13}C, and ^{1}H-^{13}C HSQC NMR study. The composition and nature of lignin phenols are also determined by compound-specific **radiocarbon** analysis (CSRA) technique. Microscopy techniques such as confocal microscopy along with Histochemical Mäule staining provides indication for S-units in composition of lignin at cellular level. The H/S/G composition can also be determined by laser capture microdissection combined with the microanalysis of lignins. Mass spectrometry (MS) techniques such as Fourier transform ion cyclotron resonance mass spectrometry (FT-ICR MS) and Time-of-flight **secondary** ion mass spectrometry (ToF-SIMS) were used for the analysis of the depolymerized fragments of lignin polymers, structural determination of **monolignols**, syringyl to guaiacyl (S/G) ratio in order to obtain information on the complex polymer structure of lignin present in plant cell walls. It was found that **rupture** of inter-unit linkages at 8-O-4′, 8-1′, 8-5′, and 8-8′ in lignin showed m/z 137 and 151 due to guaiacyl ring. FT-ICR MS of wheat straw lignin showed some regularity with a difference of 44.026 m/z (C_2H_4O) units 100 suggesting lignin is not a completely random polymer. Other MS technique such as jet-cooled thermal desorption molecular **beam** (TDMB), secondary ion MS (SIMS), synchrotron vacuum-ultraviolet secondary neutral MS (VUVSNMS) were also used to understand the fragmentation mechanism of monolignols under different energetic processes. The positive ion SIMS spectrum of coniferyl alcohol showed characteristic peaks at m/z 137 and 151. A study on wheat straw lignin using atmospheric pressure photoionization quadrupole time-of-flight mass spectrometry (APPI-QqTOF-MS) provided evidence that grass lignin is composed of repeating phenylcoumaran units, which are formed from two di-coniferyl units linked by the C8 − C′5 covalent bond and the ether C7-O-4′ linkage, forming a furan-like ring attached to an aromatic coumaran ring. Py-MBMS of grass bagasse gave a distinctive fragmentation pattern with high m/z 114 consistent with expected xylan enrichment, and fragments at m/z 150 and 120 indicated coumaryl derivatives, presumably from the **hydroxycinnamic acid** groups, PCA and FA. Simple techniques such as GC-MS analysis of **pyrolysed** softwood and hardwood sample confirmed the presence of syringyl and guaiacyl groups in hardwood and softwood lignin. Non-wood fibers such as hemp, flax, jute, sisal and abaca, alkali lignins have been analyzed. Hemp and flax have low S/G ratios, while jute, sisal and abaca showed high S/G ratios, as revealed by Py-GC/MS and FTIR analysis. Py/TMAH showed a significant amount of PCA in the abaca lignin and much lower cinnamic contents in the other lignins. This analysis also confirmed that PCA is attached to cell walls through ester bonds, while FA through ether-linkage except sisal where, the linkages are found to be in reverse order. Softwood lignin pyrolysis afforded coniferyl derivatives while hardwood lignin coniferyl and sinapyl derivatives and grass lignin p-vinylphenol as confirmed by GCMS studies. Thermal characterisation such as DSC and TGA, and thermorheological analysis can also provide insight about nature of wood. The yield of carbon generated from SKL and soda hardwood lignin was found to be 37% and 34% as analyzed by TGA studies at 900 ℃. The **viscosity** of softwood and hardwood lignin was found to be considerably different due to their different chemical structures and molecular weights and former shows a lower value 2.8 poise while later has 3.5 poise at 1.8 s^{-1} at 225℃.

Selected from: *RSC Adv.*, 2014, 4: 21712-21752

Words and Expressions:

structural configuration [ˈstrʌktʃərəl] [kənˌfɪɡjəˈreɪʃn] 结构形式，结构布局
quantification [ˌkwɑːntɪfɪˈkeɪʃn] n. [统计]定量，量化
correlated [ˈkɔrəˌletɪd] adj. 有相互关系的
predominance [prɪˈdɑːmɪnəns] n. 优势；卓越
cellulolytic enzyme lignin [ˌsɛljʊloˈlɪtɪk] [ˈenzaɪm] [ˈlɪɡnɪn] 木质素纤维素分解酶
radiocarbon [ˌreɪdioʊˈkɑːrbən] n. [核]放射性碳
secondary [ˈsekənderi] adj. 第二的；中等的；次要的；中级的
monolignols [ˈmɑːnoʊlɪɡnɔls] 木质素单体
rupture [ˈrʌptʃər] n. 破裂；决裂
beam [biːm] n. 横梁；光线；电波；船宽
hydroxycinnamic acid 羟基肉桂酸
pyrolyzed [ˈpaɪrəˌlaɪz] vt. 使……热解；使……裂解
viscosity [vɪˈskɑːsəti] n. 黏性；黏度

Notes:

1) Non-woody biomass such as corn stover was studied for the structural changes showed by lignin and LCCs characterized by alkaline nitrobenzene oxidation, ^{13}C, and ^{1}H-^{13}C HSQC NMR study.

以玉米秸秆等非木质生物质为研究对象，使用^{13}C、^{1}H-^{13}C HSQC等核磁共振为手段研究其木质素和LCCs复合体在碱性硝基苯氧化下表现出的结构变化。

2) Softwood lignin pyrolysis afforded coniferyl derivatives while hardwood lignin coniferyl and sinapyl derivatives and grass lignin p-vinylphenol as confirmed by GCMS studies.

GC-MS 研究表明软木木质素热解制得松柏醇类衍生物，硬木木质素热解可制得松柏与芥子醇类衍生物，而草类木质素热解制得对乙烯苯酚类化合物。

Recommended Reading Materials:

1. Balogun A O, Lasode O A, McDonald A G. Thermo-analytical and physico-chemical characterization of woody and non-woody biomass from an agro-ecological zone in Nigeria. *BioResources*, 2014, 9(3): 5099-5113.

2. Li H, McDonald A G. Fractionation and characterization of industrial lignins. *Industrial Crops and Products*, 2014, 62: 67-76.

Lesson 19 Depolymerization of Lignin

There are generally two main **approaches** for lignin **depolymerization**: **pyrolysis**, oxidation and **hydrotreating** (hydrogenolysis, **deoxygenation**). In some cases, enzymatic depolymerization methods have also been described. However, depolymerisation processes are often not too well understood in terms of **mechanism**. When analyzing current general **methodologies** and recent **protocols**, there are two main considerations that should be taken into account for lignin valorisation **purposes**: (i) maximisation of the activity and stability of the catalyst under the chosen conditions (as mild as possible), bearing in mind the **bulky** nature of the biomass source and the reaction conditions, and (ii) more importantly, the repolymerisation and self-condensation capability of lignin under processing conditions (due to the formation of radicals and/or C-C bond forming self-condensation reactions in acidic media) which eventually leads to a complex pool of re-condensed aromatics.

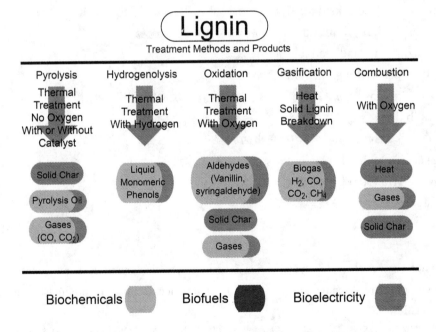

Figure 19.1 Lignin treatment methods and products

In other cases, when depolymerisation is achieved under generally harsh reaction conditions (high hydrogen pressures, temperatures > 440 ℃), it produces a range of products composed mainly of simple aromatics. Another important aspect to consider for lignin depolymerisation is the addition of hydrogen to the products. In some cases, reduction caused by hydrogen is not avoided and cannot be controlled. This hydrogenation side reaction becomes more **prominent** in the presence of a metal in the system as it could act as a surface for the reduction. It is critical that the impact of these issues on product quality and hydrogen economy be addressed to obtain a clean valorization strategy for lignin. Previous work and results in this field point to a general consensus that catalytic **hydrogenolysis** is a valuable approach to unravel many structural insights into lignin through the detection of the identity of the lignin degradation products. In any case, the design of an active and stable system able to work under mild reaction conditions will offer a significant step forward in lignin deconstruction **strategies**. Chemical and enzymatic depolymerisation strategies have been developed in recent years aiming to recover the **aromatic monomers** in lignin. Relevant literature examples mostly related to hydrotreating technologies that advance the understanding of lignin deconstruction will be subsequently discussed in the next sections. While thermochemical (e.g. pyrolysis) and oxidative protocols can contribute to lignin depolymerisation, these will not be discussed in detail in this contribution.

Selected from: *Chem. Soc. Rev.*, 2014, 43: 7485-7500.

Words and Expressions:

approach [əˈprəʊtʃ] n. 方法；途径；接近
depolymerization [ˌdiːˌpɒlɪməraɪˈzeɪʃən] n. [高分子]解聚(合)作用
pyrolysis [paɪˈrɒlɪsɪs] n. [化学]热解；[化学]高温分解
hydrotreat [ˈhaɪdrətriːt] vt. 氢化处理；加氢精制
deoxygenation [diːˌɔ-ksidʒiˈneɪʃən] n. [化学]脱氧；[化学]脱氧作用
mechanism [ˈmek(ə)nɪz(ə)m] n. 机制；原理，途径
methodology [ˌmɛθədˈɑlədʒi] n. 方法论
protocol [ˈprəʊtəkɒl] n. 协议；草案；礼仪
purpose [ˈpɜːpəs] n. 目的；用途；意志；vt. 决心；企图；打算
bulky [ˈbʌlkɪ] adj. 体积大的；庞大的；笨重的
prominent [ˈprɒmɪnənt] adj. 突出的，显著的；杰出的；卓越的
hydrogenolysis [ˌhaɪdrɒdʒəˈnɑlɪsɪs] n. 氢解作用
strategy [ˈstrætədʒɪ] n. 战略，策略
aromatic monomer 芳香族单体

Notes:

1) In other cases, when depolymerisation is achieved under generally harsh reaction conditions (high hydrogen pressures, temperatures >440℃), it produces a range of products composed mainly of simple aromatics.

在其他情况下，解聚反应普遍是在苛刻的反应条件下完成的(高氢气压力，温度>440℃)，会产生一系列主要由简单芳烃组成的产品。

2) In any case, the design of an active and stable system able to work under mild reaction conditions will offer a significant step forward in lignin deconstruction strategies.

无论如何，设计一种高活性且稳定并能够在温和反应条件进行反应的体系，将极大地推进木质素解构研究策略。

Recommended Reading Materials:

1. Das A, Rahimi A, Ulbrich A, Alherech M, Motagamwala A H, Bhalla A, da Costa Sousa L, Balan V, Dumesic J A, Hegg E L, Dale B E, Ralph J, Coon J J, Stahl S S. Lignin conversion to low-molecular-weight aromatics via an aerobic oxidation-hydrolysis sequence: comparison of different lignin sources. *ACS Sustainable Chem. Eng.*, 2018, 6: 3367-3374.

2. Salvachúa D, Karp E M, Nimlos C T, Vardon D R, Beckham G T. Towards lignin consolidated bioprocessing: simultaneous lignin depolymerization and product generation by bacteria. *Green Chem.*, 2015, 17: 4951-4967.

Lesson 20 Fast Pyrolysis of Lignin

Lignin is an important cell wall component of biomass, especially the woody species.

Recently Laskar *et al.* and Saidi *et al.* reviewed the **pathway** for lignin conversion with the focus on lignin isolation and catalytic hydrodeoxygenation of lignin-derived bio-oils. The three basic structural units of lignin are *p*-coumaryl alcohol, coniferyl alcohol, and sinapyl alcohol. The relative abundances of *p*-coumaryl alcohol, coniferyl alcohol, and sinapyl alcohol units vary with the sources of biomass but the **linkages** are similar. Among all the interphenylpropane linkages involved in lignin substructures, the guaiacylglycerol β-arylether substructure is the most abundant (40%-60%). The abundances of other substructures found in lignin macromolecules are phenyl coumarone (10%), diarylpropane (5%-10%), pinoresinol (5% or less), biphenyl (5%-10%), and diphenyl ether (5%). Lignin has a high **resistance** to **microbial** and chemical attacks due to its complex three-dimensional network formed by different non-phenolic phenylpropanoid units linked with a variety of ether and C-C bonds, and is the most recalcitrant component of lignocellulose.

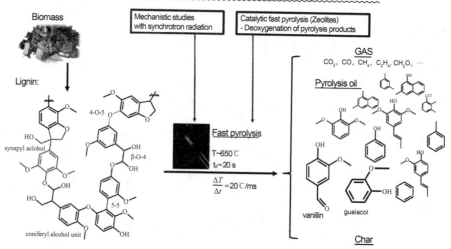

Figure 20.1 Graphical overviews of lignin structure and fast pyrolysis products

Thermal pyrolysis can break down these phenyl-propane units of the macromolecule **lattice**. The pyrolysis of lignin starts with dehydration at about 200℃ followed by breakdown of the β-O-4 linkage, leading to the formation of guaiacol, dimethoxyphenol, dimethoxyacetophenone (DMAP), and trimethoxyacetophenone (TMAP). The β-O-4 bond **scission** occurs at temperatures between 250℃ and 350℃. α-, and β-aryl-alkyl-ether linkages break down between 150℃ and 300℃. The **aliphatic** side chains also start **splitting off** from the aromatic ring at about 300℃. An even higher temperature (370-400℃) is required to break the C-C bond between lignin structural units. More generally, there are three kinds of bond cleavage including two C-O bond cleavages and one side chain C-C bond cleavage. The cleavage of a methyl C-O bond to form two-oxygen-atom products is the first reaction to occur in the **thermolysis** of 4-alkylguaiacol at 327-377℃. Then the cleavage of the aromatic C-O bond leads to the formation of one-oxygenatom products. The side chain C-C bond cleavage occurs between the aromatic ring and a-carbon atom. However, the product distribution varies with the source of biomass. Guaiacol is the main product from coniferous wood while guaiacol and pyrogallol **dimethyl ether** are dominant from deciduous woods. Lignin produces more char and tar than wood despite the higher **methoxyl** content of lignin.

Selected from: *Chem. Soc. Rev.*, 2014, 43: 7594-7623.

Words and Expressions:

  pathway ['pɑːθweɪ] *n.* 路，道；途径，路径
  linkage ['lɪŋkɪdʒ] *n.* 连接；结合；联接
  resistance [rɪ'zɪst(ə)ns] *n.* 阻力；电阻；抵抗；反抗；抵抗力
  microbial [maɪ'krəʊbɪəl] *adj.* 微生物的；由细菌引起的
  thermal ['θɜːm(ə)l] *adj.* 热的；热量的；保热的；*n.* 上升的热气流
  lattice ['lætɪs] *n.* [晶体]晶格；格子；格架
  scission ['sɪʃ(ə)n] *n.* 切断，分离；断开
  aliphatic [ˌælɪ'fætɪk] *adj.* 脂肪质的，[有化]脂肪族的
  split off 分离；分裂
  thermolysis [θə'mɒlɪsɪs] *n.* [化学]热解；散热作用
  dimethyl ether [有化]二甲醚；甲醚
  methoxyl [mɪ'θɒksʌɪl] *n.* 甲氧基

Notes:

1) Lignin has a high resistance to microbial and chemical attacks due to its complex three-dimensional network formed by different phenolic phenylpropanoid units linked with a variety of ether and C-C bonds, and is the most recalcitrant component of lignocellulose.

木质素是木质纤维素中最顽强的成分，它由不同的酚型苯丙烷单元通过醚键和 C-C 键连接而形成复杂的三维网状结构，对微生物和化学攻击具有很高的抵抗力。

2) The pyrolysis of lignin starts with dehydration at about 200℃ followed by breakdown of the β-O-4 linkage, leading to the formation of guaiacol, dimethoxyphenol, dimethoxyacetophenone (DMAP), and trimethoxyacetophenone (TMAP).

木质素在 200℃ 左右开始热解脱水，然后 β-O-4 键断裂，形成愈创木酚、二甲氧基苯酚、二甲氧基苯乙酮(DMAP)及三甲氧基苯乙酮(TMAP)。

Recommended Reading Materials:

1. Chu S, Subrahmanyam A V, Huber G W. The pyrolysis chemistry of a β-O-4 type oligomeric lignin model compound. *Green Chem.*, 2013, 15, 125-136.

2. Patwardhan P R, Brown R C, Shanks B H. Understanding the fast pyrolysis of lignin. *ChemSusChem.*, 2011, 4, 1629-1636.

Lesson 21   Association of Lignin

In 1979, Lindström studied **association** and **precipitation** of kraft lignin (Indulin ATR, 11037-2, Westvaco Co., Charleston, SC., USA) in aqueous solutions varying in pH from 8.6 to 3.7. It was found that the degree of association between lignin molecules increased with decreasing ionization of the **carboxylic group** in the lignin. They suggested that the hydrogen bonding between carboxylic groups and various ether oxygens and hydroxylic groups was responsible for inter-molecular association of lignin. Later, McCarthy *et al.* studied association of kraft lignin, isolated from the black liquor of Douglas fir, in aqueous NaOH solutions.

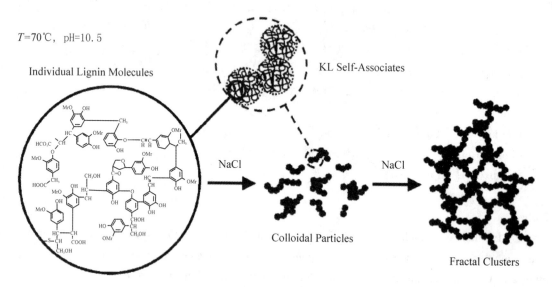

Figure 21.1   Schematic illustration of association of lignin

Lignin molecules adopted an expanded **random** coil conformation in dilute alkaline solutions at pH 9.5 as determined by the relationship between **radius** of **gyration** and molecular weight. Association between lignin molecules increased with increasing lignin concentration under aqueous conditions of pH 13 and 14 (Figure 21.1). Hydrogen bonding was not considered as the driving force for association of lignin molecules at higher concentrations because acetylation of this lignin samples did not appreciably affect the **proportion** of higher molecular weight complexes in N, N-dimethylformamide. Hydrophobic interactions alone were believed to be not strong enough to overcome the **repulsion** between charged lignin molecules. It was thus proposed that non-bonded orbital interactions between aromatic moieties of the lignin molecules were responsible for the association. More recently, Qiu *et al.* **investigated** the mechanism of kraft lignin association in tetrahydrofuran (THF) using iodine as a probe. The results suggested that the non-bonded **orbital** interactions (π-π interactions) of the aromatic groups caused the association of lignin molecules. Zhou et al. reported association of sodium lignosulfonate in water with lignin concentrations varying from 10 to 2 500 mg · L^{-1}. The analysis suggested that individual lignin molecules started to associate to form **hollow micelles** which further **clustered** into larger **loose aggregates** with increasing lignin concentration. It was proposed that lignin molecules associated via hydrogen bonding and intermolecular van der Waals forces. The **electrostatic repulsion** between lignin molecules made the aggregates rather loose. Vainio et al. discovered that the lignosulfonate molecules assumed a flat shape and the charged groups were mostly located at the flat surfaces. Lignosulfonate molecules were shown to aggregate along the edges of the **flat particles**.

Selected from: Green Chem., 2016, 18: 5693-5700.

Words and Expressions:

association [əˌsəʊʃi'eɪʃn] n. 协会，联盟，社团；联合；联想
precipitation [prɪˌsɪpɪ'teɪʃ(ə)n] n. [化学]沉淀，[化学]沉淀物
carboxylic group 羧基
random ['rændəm] adj. 随机的；任意的；胡乱的
radius ['reɪdɪəs] n. 半径，半径范围
gyration [dʒaɪ'reɪʃ(ə)n] n. 旋转，[力]回转；螺层
proportion [prə'pɔːʃ(ə)n] n. 比例，占比；
repulsion [rɪ'pʌlʃ(ə)n] n. 排斥；反驳；反感；厌恶
investigated [ɪn'vestɪgeɪt] v. 调查；研究
orbital ['ɔːbɪt(ə)l] adj. 轨道的；眼窝的
micelles [maɪ'sɛlz; mɪ'sɛlz] n. [化学]胶团；微粒；[分子生物]微胶粒
clustered ['klʌstəd] adj. 成群的；聚集成群的；聚合的
loose aggregates ['ægrɪgət；(for v.) 'ægrɪgeɪt] 松散聚集体
electrostatic repulsion 静电排斥
flat particles 平面粒子

Notes:

1) Hydrogen bonding was not considered as the driving force for association of lignin molecules at higher concentrations because acetylation of this lignin samples did not appreciably affect the proportion of higher molecular weight complexes in N, N-dimethylformamide.

由于该木质素样品的乙酰化对 N, N-二甲基甲酰胺中高相对分子质量配合物的比例没有明显影响，因此氢键不是高浓度木质素分子结合的驱动力。

2) The analysis suggested that individual lignin molecules started to associate to form hollow micelles which further clustered into larger loose aggregates with increasing lignin concentration.

分析表明随着木质素浓度的增加，单个木质素分子开始结合形成空心胶束，并进一步聚集成更大的松散聚集体。

Recommended Reading Materials:

1. Contreras S, Gaspar A R, Guerra A, Lucia L A, Argyropoulos D S. Propensity of lignin to associate: light scattering photometry study with native lignins. *Biomacromolecules*, 2008, 9: 3362-3369.

2. Deng Y, Feng X, Zhou M, Qian Y, Yu H, Qiu X. Investigation of aggregation and assembly of alkali lignin using iodine as a probe. *Biomacromolecules*, 2011, 12: 1116-1125.

Lesson 22 Reactivity of Lignin in Ionic Liquids

Virtually all pretreatment options induce chemical changes in both hemicellulose and lignin, for example polymer **fragmentation**, chemical **transformation** or functionalization, while the cellulose usually remains largely chemically unchanged (NB: structural modifications of cellulose have been observed). Deconstruction efforts involving ionic liquids are conducted at elevated temperature; therefore such effects need to be considered for ionic liquid treatment of lignocellulose.

A number of observations regarding the modification of lignin during ionic liquid deconstruction have been published. Tan et al. reported that lignin extracted with [C_2C_1im][ABS] has a lower **molecular weight** and a narrower **polydispersity** than a lignin obtained by aqueous auto-catalysed pretreatment. George et al. studied the impact of a range of ionic liquids on several commercial lignins and demonstrated a profound anion effect on the fragmentation mechanism and the degree of polymerisation, with liquids containing alkyl sulfate anions having the greatest ability to fragment the lignins and reduce polymer length. The order of molecular weight reduction was sulfates > lactate > acetate > chloride > phosphates. The functional group of the anion determined the effect rather than its size, while the cation did not play a significant role. The authors suggest that the more active anions act as nucleophiles during lignin depolymerisation. In support of this, an increased **sulfur** content of the lignin after treatment with ionic liquids with sulfur containing anions such as sulfonates and sulfamates (e.g. [ABS]$^-$ and acesulfamate) and sulfates ([$MeSO_4$]$^-$ and [HSO_4]$^-$) has been reported. This may also explain why an attempt to design a 'lignin friendly' cation by adding an aromatic side chain had only moderate success.

Analysis of the **residual** lignin in [C_2C_1im][$MeCO_2$] pretreated maple wood with 2-dimensional NMR revealed a decrease of the β-O-4 aryl ether bond content as well as deacetylation of xylan. The effect of two lignocellulose-dissolving ionic liquids on a lignin model compound featuring a β-O-4 **aryl ether linkage** has been studied. The compound was dissolved in [C_2C_1im]Cl or [C_2C_1im][$MeCO_2$] at 120 ℃. An α, β-dehydration reaction was reported for both ionic liquids. The **dehydration** was significantly faster in [C_2C_1im][$MeCO_2$] than in [C_2C_1im]Cl, possibly reflecting the acetate's greater basicity and affinity towards water. The occurrence of such dehydration reactions has also been suggested by George et al. more recently.

Cleavage of the β-O-4 aryl ether bond in guaiacylglycerol-β-guaiacyl ether by hydrogen-bond acidic monoalkylimidazolium ionic liquids at 110–150 ℃ was investigated. Differences in reactivity depending on the anion were observed. More strongly hydrogen bond-basic anions (Cl^-, Br^- and $[HSO_4]^-$) resulted in higher yields of the **cleavage** products than weakly basic anions, as well as in different routes. The enol ether product was observed as an intermediate when the ionic liquid contained a coordinating anion. Less coordinating anions such as $[BF_4]^-$ resulted in elimination of the γ hydroxyl methylene group, resulting in formation of formaldehyde.

Selected from: *Green Chemistry*, 2013, 15: 550-583.

Words and Expressions:

  fragmentation [ˌfrægmenˈteɪʃn] n. 破碎；分裂
  transformation [ˌtrænsfərˈmeɪʃn] n. [遗] 转化；转换；改革；变形
  molecular weight [məˈlekjələr][weɪt] [化学] 分子量
  polydispersity [ˌpɒlɪdɪsˈpɜːsɪtɪ] n. [化学] 多分散性
  sulfur [ˈsʌlfɚ] n. 硫，硫黄；硫黄色
  residual [rɪˈzɪdʒuəl] adj. (数量)剩余的；n. 剩余物，残渣
  aryl ether linkage [ˈærɪl][ˈiːθər][ˈlɪŋkɪdʒ] 芳基醚键
  dehydration [ˌdiːhaɪˈdreɪʃn] n. 脱水
  cleavage [ˈkliːvɪdʒ] n. 劈开，分裂

Notes:

1) George *et al*. studied the impact of a range of ionic liquids on several commercial lignins and demonstrated a profound anion effect on the fragmentation mechanism and the degree of polymerisation, with liquids containing alkyl sulfate anions having the greatest ability to fragment the lignins and reduce polymer length.

乔治等人研究了一系列离子液体对几种市售木质素的影响，并证明了阴离子对木质素裂解机理和聚合度有深远的影响，其中含有烷基硫酸盐阴离子的离子液体对木质素的裂解与解聚拥有最佳的效果。

2) The effect of two lignocellulose-dissolving ionic liquids on a lignin model compound featuring a β-O-4 aryl ether linkage has been studied.

研究了两种可溶解木质纤维素的离子液体对具有 β-O-4 芳基醚键的木质素模型化合物的影响。

Recommended Reading Materials:

1. Jia S, Cox B J, Guo X, Zhang Z C, Ekerdt J G. Hydrolytic cleavage of β-O-4 ether bonds of lignin model compounds in an ionic liquid with metal chlorides. *Industrial & Engineering Chemistry Research*, 2011, 50(2): 849-855.

2. Zavrel M, Bross D, Funke M, Büchs J, Spiess A C. High-throughput screening for ionic liquids dissolving (ligno-) cellulose. *Bioresource technology*, 2009, 100(9): 2580-2587.

Lesson 23 Functionalization of Lignin Hydroxyl Groups (I)

Lignin presents in its structure phenolic hydroxyl groups and aliphatic hydroxyl groups at the C-γ and C-α positions on the side chain. Phenolic hydroxyl groups are the most reactive functional groups and can affect the chemical reactivity of the newly formed material. Modifications on hydroxyl groups can lead to the formation of polyol derivatives of lignin. For that to occur, several reactions have been studied to functionalize the lignin with different functional groups, and these include reactions such as **alkylation, esterification, etherification, phenolation and urethanization**, which are briefly discussed in the following part.

a) Alkylation/dealkylation

Lignin presents different sites for alkylation, including the oxygen atoms of the hydroxyl, carbonyl and carboxyl groups. There are three methods for alkylating lignin, which can include a reaction with diazoalkanes, a reaction with alcohol in the presence of a catalyst (e.g., hydrochloric acid), or a reaction using alkyl sulfates and sodium hydroxide. For example, demethylation is one of the most well-known examples of alkylation/dealkylation reactions that involve lignin in which the demethylated lignin structures are a byproduct in DMSO production. The synthesis starts with the reaction between lignin and molten sulfur in alkaline media. This **demethylated** lignin, in combination with polyethylenimine, is used as a formaldehyde-free wood adhesive.

b) Esterification

Esterification is possibly the easiest of the reactions to produce lignin-based polyesters that involve the hydroxyl groups of lignin. Esterification can be performed by using three different procedures, which are: ring opening reactions using cyclic esters, condensation polymerization with carboxylic acid chloride, and **dehydration** polymerization with dicarboxylic acids. For example, the reaction between e-caprolactone with hydroxypropylated lignin produces lignin-based polyesters with a star-like shape. In these star shapes, hydroxypropylated lignin forms the core, whereas the polycaprolactones form the arm segments. Apart from e-caprolactone, lactide can also be used for copolymerization with lignin to produce **lignin-polylactide polyester** using triazabicyclodecene as a catalyst in metal and solvent free systems. An example of condensation polymerization was obtained when lignin-based polyesters were prepared by reacting lignin with the dicarboxylic acid chlorides, sebacoyl chloride and terephtaloyl chloride in organic solvents. Dehydration polymerization procedures that use dicarboxylic acids have been proposed. For example, dimeric acid, a dicarboxylic acid that is synthesized by dimerizing unsaturated fatty acids obtained natural oils, can be applied in the co-esterification with enzymatically hydrolyzed lignin, which makes the co-ester more flexible. Additionally, adding carboxy-telechelic polybutadiene to lignin can also produce lignin-based polyesters. These lignin esterification reactions that involve the e-caprolactone and different anhydrides are performed in order to increase the lignin reactivity for the production of lignin-based epoxy resins, polyurethanes and unsaturated **thermosetting composites**.

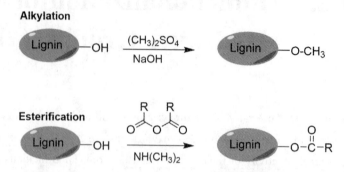

Selected from: *Progress in Materials Science*, 2018, 93: 233-269.

Words and Expressions:

    alkylation [ˌælkəˈleʃən] n. [有化] 烃化；烷化
    esterification [eˌsterəfəˈkeɪˌʃən] n. 酯化(作用)
    etherification [iˌθe-rifiˈkeiʃən] n. [有化] 醚化；醚化作用
    phenolation [ˈfiːnɔːleiʃən] n. 酚化
    urethanization 氨基甲酸酯化
    dehydration [ˌdiːhaɪˈdreɪʃn] n. 脱水
    lignin-polylactide polyester 木质素-聚乳酸聚酯
    thermosetting composites [ˈθɜ·moˌsɛtɪŋ] [kəmˈpɑzɪts] 热固性复合材料

Notes:

1) Phenolic hydroxyl groups are the most reactive functional groups and can affect the chemical reactivity of the newly formed material.

酚羟基是最活泼的官能团，可影响新形成材料的化学反应性。

2) For example, dimeric acid, a dicarboxylic acid that is synthesized by dimerizing unsaturated fatty acids obtained natural oils, can be applied in the co-esterification with enzymatically hydrolyzed lignin, which makes the co-ester more flexible.

例如，二聚酸是通过将天然油中的不饱和脂肪酸二聚而合成的二羧酸，可用于酶解木质素的共酯化反应中，从而使共聚酯更具柔性。

Recommended Reading Materials:

    1. Fulcrand H, Dueñas M, Salas E, Cheynier V. Phenolic reactions during winemaking and aging. *American Journal of Enology and Viticulture*, 2006, 57(3): 289-297.

    2. Gellerstedt G, Lindfors E. Structural changes in lignin during kraft cooking. Part 4. Phenolic hydroxyl groups in wood and kraft pulps. *Svensk Papperstidn*, 1984, 87(15): R115-R118.

Lesson 24 Functionalization of Lignin Hydroxyl Groups (Ⅱ)

c) Etherification

The preparation of lignin-based **polyethers** can be performed by one or any combination of the following procedures: polymerization using the alkylene oxides (e. g., ethylene oxide and **propylene oxide**), polymerization with epichlorohydrin, cross-linking by using diglycidyl ethers, and solvolysis of lignin with ethylene glycol. The aromatic moieties present in the lignin structure will improve the mechanical and thermal properties of epoxy resins. The oxypropylation is the most used etherification method to modify lignin using propylene oxide in the presence of an alkaline solution, in order to prepare lignin-based epoxy resins. The resulting solution was treated with epichlorohydrin and cured using m-phenylene diamine for cross-linking. Lignosulfonate is first submitted to phenolation using phenol, b-naphtol and bisphenol, and then **epoxidized** using epichlorohydrin in order to improve the reactivity of the lignin and increase the phenolic group content. Lignin from woody biomass can also be grafted with ethylene glycol/ethylene carbonate by solvolysis at 150 ℃ to introduce ethylene glycol chains into the hydroxyl groups of lignin.

d) Phenolation

Phenolation, also known as phenolysis, is the process by which lignin is modified by reaction with phenol in the presence of organic solvents such as methanol or ethanol in an acidic medium. This reaction is commonly used to modify lignosulfonates in order to increase the content of phenol groups and improve the reactivity of the target lignin structure. Phenolysis is utilized in the synthesis of **phenol-formaldehyde resins** before the condensation reaction with formaldehyde. The reaction can be performed at 70 ℃ for a few hours, after which lignin is added to a phenol-ethanol solution. The curing time and viscosity of the lignin phenol formaldehyde resins are comparable to standard commercial phenol-formaldehyde resins, and this product is used to provide wood adhesive capacity in the construction of particle board. Additionally, lignin that has been modified with **cardanol**, a natural phenol, can also be used to produce lignin-based polyurethane films with improved properties such as film flexibility.

e) Urethanization

The urethanization process involves the reaction between lignin hydroxyl groups and isocyanate groups to form a **urethane** link. Traditionally, polyurethanes have been produced from polyols and diisocyanates to provide versatile products. Examples of such products include low temperature **elastomers** and flexible or rigid adhesives with high tensile strength. Since the lignin structure is rich in hydroxyl groups, it can function as a polyol. The improvement of the mechanical properties of lignin-based polyurethanes can be achieved by chemical modifications on lignin such as hydroxyalkylation to introduce **soft segments**, or by adding other polyols such as polyethylene glycol (PEG) or other diols. There are two different approaches for the synthesis of lignin-based polyurethane. The first approach is a one-step reaction that occurs by adding diisocyanate and another by adding diol as the co-monomer. The second approach is a two-step reaction with the first step being the production of a prepolymer using isocyanate together with a polyol, and the second the polymerization of lignin with the prepolymer, acting as a chain extender. In the first approach, lignin is directly used in combination with **isocyanate** and polyols without further chemical modification in order to produce lignin-based polyurethane. For instance, lignin can be mixed with 4, 40-diphenylmethane diisocyanate and PEG to produce lignin-based polyurethane. An example of the second approach, is the formation of prepolymers with hydroxyl terminated polybutadiene and 2, 4-toluene diisocyanate, which are subsequently reacted with lignin to produce a **polyurethane**.

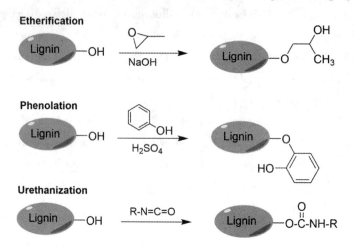

Selected from: *Progress in Materials Science*, 2018, 93: 233-269.

Words and Expressions:

    polyethers ['pɑliˌiθɚs] [高分子]聚醚
    propylene oxide ['propəlin]['ɑːksaɪd] [有化]氧化丙烯
    epoxidized *adj.* 使环氧化的 *v.* 使环氧化(epoxidize 的过去分词)
    phenol-formaldehyde resins ['fiːnɔːl][fɔːrˈmældɪhaɪd]['rɛzɪnz] 酚醛树脂
    cardanol 腰果酚
    urethane [jʊˈrɛθen] 氨基甲酸乙酯,尿烷
    elastomers [ɪˈlæstəmərs] *n.* [力]弹性体(elastomer 的复数)
    soft segments [sɔːft]['sɛgmənt] 软段
    isocyanate [ˌaɪsoˈsaɪənət] *n.* [无化]异氰酸盐
    polyurethane [ˌpɑːliˈjʊrəθeɪn] *n.* 聚氨酯

Notes:

1) Lignin from woody biomass can also be grafted with ethylene glycol/ethylene carbonate by solvolysis at 150 ℃ to introduce ethylene glycol chains into the hydroxyl groups of lignin.

木质生物质中的木质素也可以在 150 ℃溶剂中分解，并与乙二醇/碳酸乙烯酯发生接枝反应，将乙二醇链引入木质素的羟基中。

2) The second approach is a two-step reaction with the first step being the production of a prepolymer using isocyanate together with a polyol, and the second the polymerization of lignin with the prepolymer, acting as a chain extender.

第二种方法是两步反应，第一步是使用异氰酸酯与多元醇一起生产预聚物，第二步是木质素与预聚物的聚合，充当扩链剂。

Recommended Reading Materials：

1. Gómez-Fernández S, Ugarte L, Calvo-Correas T, Peña-Rodríguez C, Corcuera M A, Eceiza A. Properties of flexible polyurethane foams containing isocyanate functionalized kraft lignin. *Industrial Crops and Products*, 2017, 100：51-64.

2. Panesar S S, Jacob S, Misra M, Mohanty A K. Functionalization of lignin：Fundamental studies on aqueous graft copolymerization with vinyl acetate. *Industrial Crops and Products*, 2013, 46：191-196.

Lesson 25　Lignin-derived Polymers

Lignin can be used for the development of lignin graft **copolymers** in which the polymer chains are attached to the hydroxyl groups on the lignin structure, which produces a star-like branched copolymer with a lignin core. Some examples of polymerization reactions were briefly mentioned above in the chemical modifications section. "Grafting from" technique would be introduced in this part. In the "grafting from" technique, lignin is used as **macro-initiator** for the polymerization during which a monomer reacts with hydroxyl groups present in lignin and the polymer chain is assembled on the lignin core. The ring opening polymerization of different monomers and radical polymerization of **vinylic monomers** are two approaches that have been performed to elaborate lignin-graft copolymers by "grafting from" procedure.

a) Ring opening polymerization

The reaction of lignin AOH groups with propylene oxide is one of the most common ring opening polymerization reactions that lead to the production of oxypropylated lignin. Using this reaction, the decrease of the Tg and viscosity of the copolymer can be achieved by increasing the grafted chain length. The resulting copolymer has been commonly used as a "macro"-monomer the synthesis of polyurethane foams. Other polymers than propylene oxide have been used for the production of lignin graft copolymers, and these include e-caprolactone and **lactide**. The reaction between the hydroxyl groups in lignin and e-caprolactone leads to the production of lignin-polycaprolactone copolymers in which the ratio of e-caprolactone/hydroxyl groups determines the length of the **grafted polycaprolactone**. The chain length strongly affects the thermal properties of the resulting copolymers

b) Radical polymerization.

The radical polymerization of vinylic monomers onto lignin has also been used to prepare lignin-graft copolymers by a "grafting from" procedure. This process involves the creation of a radical on the lignin structure that initiates the polymerization of a vinylic monomer, usually by using irradiation or a chemical initiator (peroxide), after which the synthesis proceeds. Several copolymers have been produced by this process, including **lignin-polystyrene**, **lignin-poly(acrylic acid)** and **lignin-poly(vinyl acetate)**.

c) Atom transfer radical polymerization

In order to overcome the abovementioned limitations, the atom transfer radical polymerization (ATRP) was developed as an efficient method to regulate the radical grafting of vinylic monomers onto the lignin structure. The ATRP method allows the formation of long polymer chains with well-defined structures and low dispersity, provided that the radical polymerization is carried out in a controlled manner. Briefly, the method starts with the esterification of lignin by using 2-bromoisobutyryl bromide as the initiator. The monomers can be grafted onto the lignin macroinitiator in the presence of a catalyst such as CuBr with 1, 1, 4, 7, 10, 10-hexamethyltriethylenetetramine as the ligand, which leads to the formation of lignin-based copolymers. By using this method, the Nisopropylacrylamide (NIPAM) was copolymerized with Kraft lignin to produce thermoresponsive lignin-polyNIPAM in which the thermal decomposition temperature of the copolymers increased with increasing of the polymerization degree.

Later, the same method was used to graft polyNIPAM in lignin nanofiber mats produced by **electrospinning**, with both thermal and ionic responsive characteristics.

Selected from: *Progress in Materials Science*, 2018, 93: 233-269.

Words and Expressions:

copolymers [kɒˈpɑləmɚs] n. [高分子] 共聚物(copolymer 的复数)
macro-initiator [ˈmækroʊ] [ɪˈnɪʃieɪtər] 大分子引发剂
vinylic monomers 乙烯单体
lactide [ˈlæktaɪd] n. 丙交酯；[有化] 交酯
grafted polycaprolactone [græftɪd] [ˌpɒliˌkæprəˈlæktəun] 接枝聚己酸内酯
lignin-polystyrene [ˈlɪgnɪn] [ˌpɑːliˈstaɪriːn] 木质素—聚苯乙烯
lignin-poly(acrylic acid) [ˈlɪgnɪn] [ˌpɑːliˈ] [əˈkrɪlɪk] [ˈæsɪd] 木质素—聚丙烯酸
lignin-poly(vinyl acetate) [ˈlɪgnɪn] [ˌpɑːli] [ˈvaɪnl] [ˈæsɪteɪt] 木质素—聚醋酸乙烯酯
electrospinning [ɪˈlektroʊˈspɪnɪŋ] 静电纺丝

Notes:

1) The reaction of lignin AOH groups with propylene oxide is one of the most common ring opening polymerization reactions that lead to the production of oxypropylated lignin.

木质素 AOH 基团与环氧丙烷的反应是制备氧丙基木质素的最常见的开环聚合反应之一。

2) The ATRP method allows the formation of long polymer chains with well-defined structures and low dispersity, provided that the radical polymerization is carried out in a controlled manner.

ATRP 聚合技术使得生成的聚合物长链段具有明确结构和低分散性，这为可控自由基聚提供了可控的控制方法。

Recommended Reading Materials:

1. Karhunen P, Rummakko P, Sipilä J, Brunow G, Kilpeläinen I. The formation of dibenzodioxocin structures by oxidative coupling. A model reaction for lignin biosynthesis. *Tetrahedron Letters*, 1995, 36(25): 4501-4504.

2. Mikulášová M, Košíková B, Alexy P, Kačík F, Urgelová E. Effect of blending lignin biopolymer on the biodegradability of polyolefin plastics. World *Journal of Microbiology and Biotechnology*, 2001, 17(6): 601-607.

Lesson 26　Biodegradation of Hemicellulose

Hemicelluloses are biodegraded to monomeric sugars and **acetic acid**. Hemicellulases are frequently classified according to their action on distinct substrates. **Xylan** is the main carbohydrate found in hemicellulose. Its complete degradation requires the **cooperative** action of a variety of hydrolytic **enzymes**. An important distinction should be made between endo-1,4-β-xylanase and xylan 1,4-β-xylosidase. The former generates **oligosaccharides** from the **cleavage** of xylan; the latter works on xylan oligosaccharides, producing xylose. In addition, hemicellulose biodegradation needs accessory enzymes such as xylan esterases, ferulic and p-coumaric esterases, α-l-arabinofuranosidases, and a-4-O-methyl glucuronosidases acting **synergistically** to efficiently hydrolyze wood xylans and mannans. In the case of O-acetyl-4-O-methylglucuronxylan, one of the most common hemicelluloses, four different enzymes are required degradation: endo-1,4-β-xylanase (endoxylanase), acetyl esterase, a-glucuronidase and β-xylosidase. The degradation of O-acetylgalactoglucomanann starts with rupture of the polymer by endomannases. Acetylglucomannan esterases remove acetylgroups, and a-galactosidases eliminate galactose residues. Finally, β-mannosidase and β-glycosidase break down the endomannases-generated oligomers β-1,4 bonds.

Xylanases, the major component of hemicellulases, have been isolated from many ecological niches where plant material is present. Due to the important biotechnological exploitations of xylanases, especially in biopulping and bleaching, many publications have appeared in recent years. The white-rot fungus *Phanerochaete chrysosporium* has been shown to produce multiple endoxylanases. Also, bacterial xylanases have been described in several aerobic species and some ruminal genera. Hydrolysis of β-glycosidic linkages is carried out by acid catalytic reactions common to all glycanases. Many **microorganisms** contain multiple loci encoding overlapping xylanolytic functions. Xylanases, like many other cellulolytic and hemicellulolytic enzymes, are highly modular in structure. They consist of either a single domain or a number of different domains, classified as catalytic and non-catalytic domains. Based on the homology of the conserved amino acids, xylanases can be grouped into two different families: family 10 (F), with relatively high molecular weight, and family 11 (G), with lower molecular weight. The catalytic domains for the two families differ in their molecular masses, net charge and isoelectric points and may play a major role in determining specificity and reactivity. Biochemically and structurally, the two families are unrelated. The release of reducing sugars from purified xylan is highly dependent on the xylanase pI. Isoelectric points for endoxylanases from various microorganisms vary from 3 to 10. Optimum temperature for xylanases from bacterial and fungal origin ranges from 40℃ to 60℃. **Fungal** xylanases are generally less thermostable than bacterial xylanases. Researchers have paid special attention to thermostable hemicellulases because of their biotechnological applications (see below). Thermophilic xylanases have been described in actinobacteria (formerly actinomycetes) such as Thermomonospora and Actinomadura. Also, a very thermostable xylanase has been isolated from the hyperthermophilic primitive bacterium *Thermotoga*. Xylanases of **thermophilic** fungi are also receiving considerable attention. As in mesophilic fungi, a multiplicity of xylanases differing in stability, catalytic efficiency, and activity on substrates has been observed. The optimal temperatures vary from 60 to 80℃ and the pI ranges from 3.7 to 9.0. This diversity of xylanase isoenzymes of different molecular masses might be to allow their diffusion into the plant cell walls.

Selected from: *Int. Microbiol*, 2002, 5: 53-63.

Words and Expressions:

acetic acid　[有化] 醋酸，[有化] 乙酸
xylan　['zaɪlæn]　n. [有化] 木聚糖
cooperative　[kəʊ'ɒpərətɪv]　adj. 合作的
enzyme　['enzaɪm]　n. [生化] 酶；酶类，酵素
oligosaccharide　[ˌɒlɪgə(ʊ)'sækəraɪd]　n. [有化] 寡糖，[有化] 低聚糖
cleavage　['kliːvɪdʒ]　n. 劈开，分裂；[晶体] 解离
synergistically　协同地
microorganism　[maɪkrəʊ'ɔːg(ə)nɪz(ə)m]　n. [微] 微生物；微小动植物
fungal　['fʌŋg(ə)l]　adj. 真菌的
thermophilic　[ˌθɜːmo'fɪlɪk]　adj. 适温的，喜温的

Notes:

1) In addition, hemicellulose biodegradation needs accessory enzymes such as xylan esterases, ferulic and *p*-coumaric esterases, *a*-l-arabinofuranosidases, and *a*-4-O-methyl glucuronosidases acting synergistically to efficiently hydrolyze wood xylans and mannans.

此外，半纤维素生物降解需要辅酶，如木聚糖酯酶、阿魏酸酯酶和对香豆素酯酶、*a*-l-阿拉伯呋喃氧化酶和 *a*-4-O-甲基葡萄糖醛酸苷酶。这些酶的协同作用可以有效水解木聚糖和甘露聚糖。

2) Based on the homology of the conserved amino acids, xylanases can be grouped into two different families: family 10 (F), with relatively high molecular weight, and family 11 (G), with lower molecular weight.

根据保守氨基酸的同源性，木聚糖酶可分为两个不同的种族：相对分子质量较高的 10 (F) 和相对分子质量较低的 11 (G)。

Recommended Reading Materials:

1. Xiao L P, Shi Z J, Bai Y Y, Wang W, Zhang X M, Sun R C. Biodegradation of lignocellulose by white-rot fungi: structural characterization of water-soluble hemicelluloses. *Bioenerg. Res.*, 2013, 6: 1154-1164.

2. Popescu C M, Popescu M C, Vasile C. Structural changes in biodegraded lime wood. *Carbohyd. polym.*, 2010, 79: 362-372.

Lesson 27 Tannins

Tannins are important commercial substances, traditionally used as **tanning** agents in the leather industry, allowing the transformation of hide into **leather**. Other uses include **adhesives** (as phenol substitutes in the formulations), medical, cosmetic, **pharmaceutical** and food industrial applications. Regarding the woodland conservation, particularly to avoid the fell of trees such as quebracho, mimosa and chestnut, **alternative** tannin sources, such as locally available **residues** or byproducts, should be selected. Possible sources include wine wastes (grapeseed), chestnut peels (from chestnut industry processing), forest **exploitation** wastes, timber, pulp and paper mills residues. Some of these biomass residues are incinerated, landfilled, used in **horticulture** or as energy source. It is however known that their use for heat/power generation have operational, economic and environmental limitations; in addition, these **options** despise the valuable chemical content of the biomaterials. The extraction of tannins from vegetable residues **constitutes** then an important contribute for their reuse and **valorization**, and for tannins sustainable production. Tannins are historically classified into two groups: hydrolysable or non-hydrolysable (condensed). On the basis of their structural characteristics, however, a **classification** into four major groups is preferable: gallotannins, ellagitannins, complex tannins, and condensed tannins (Figure 27.1). Hydrolysable tannins can be fractionated into simple components, by treatment with hot water or by enzymes.

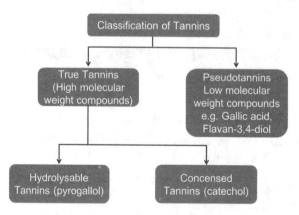

Figure 27.1 Class ification of Tannins

Condensed tannins have a higher activity towards aldehyde, being chemically and economically more interesting for the preparation of **resins**, adhesives and other applications apart from the leather. Figure 27.1 presents the structures of a catechin unit and a condensed tannin. The presence of phenolic groups in tannins clearly indicates its **anionic** nature. Phenolic groups act as weak acids, easily deprotonate, being good hydrogen donors, to form phenoxide ion which is resonance-stabilized.

There are no universal conditions for extracting tannins from vegetable sources. The yield and the composition of extracts depends on the source, type of solvent, extraction time, temperature, solid-liquid ratio, and preparation of the sample, which is commonly milled, used in fresh, frozen or dried state. The extraction procedure should be optimized in a case-specific basis. The polar nature of water makes it possible for use as an extraction solvent for many compounds. The traditional industrial method for tannin extraction from vegetable matter is exactly based on hot/boiling water, with temperatures ranging from 50℃ to 110℃, using autoclaves working in counter-current, contact times of several hours (6-10 h) and water/wood ratios equal to 2-2.4 in mass. Tannins are then concentrated, by evaporation under vacuum, to limit the oxidation.

Selected from: *Chem. Eng. J.*, 2016, 303: 575-587.

Words and Expressions:

tanning ['tænɪŋ] n. 制革；制革法；皮肤晒成褐色
leather ['leðə] n. 皮革；皮革制品；vt. 用皮革包盖；抽打；adj. 皮的；皮革制的
adhesive [əd'hiːsɪv] n. 黏合剂
pharmaceutical [ˌfɑːmə'suːtɪk(ə)l; -'sjuː-] adj. 制药(学)的；n. 药物
alternative [ɔːl'tɜːnətɪv; ɒl-] adj. 供选择的；选择性的；交替的；n. 二中择一；供替代的选择
residue ['rezəˌdjʊ] n. 剩余物；余数；筛留物
exploitation [ˌeksplɒɪ'teɪʃ(ə)n] n. 开发，开采；利用；广告推销；剥削
horticulture ['hɔːtɪˌkʌltʃə] n. 园艺，园艺学
option ['ɒpʃnz] n. 选择；期权；选择项
constitute ['kɒnstɪtjuːt] vt. 组成，构成；建立；任命
valorization [ˌvæləraɪ'zeɪʃən] n. 稳定物价
classification [ˌklæsɪfɪ'keɪʃ(ə)n] n. 分类；类别，等级
resin ['rezɪn] n. [树脂]树脂；v. 用树脂处理
anionic [ˌænaɪ'ɒnɪk] adj. 阴离子的，带负电荷的离子的
autoclaves ['ɔːtə(ʊ)kleɪv] n. 高压灭菌器；高压釜；高压锅 v. 用高压锅烹饪；用高压灭菌器消毒；用蒸压釜养护

Notes:

1) Regarding the woodland conservation, particularly to avoid the fell of trees such as quebracho, mimosa and chestnut, alternative tannin sources, such as locally available residues or byproducts, should be selected.

为了林地的养护，特别是为了避免像白坚木、含羞草和栗树等树木的砍伐，应选择其他鞣质来源，例如当地可得到的残留物或副产品。

2) The yield and the composition of extracts depends on the source, type of solvent, extraction time, temperature, solid-liquid ratio, and preparation of the sample, which is commonly milled, used in fresh, frozen or dried state.

提取液的产量和组成取决于提取液的原料、溶剂类型、提取时间、温度、固液比和来源

于研磨、新鲜、冷冻或干燥状态下等不同方法制备的样品。

Recommended Reading Materials:

1. Arbenz A, Avérous L. Chemical modification of tannins to elaborate aromatic biobased macromolecular architectures. *Green Chem.* , 2015, 17: 2626-2646.

2. Khanbabaee K, van Ree T. Tannins: classification and definition. *Nat. Prod. Rep.* , 2001, 18: 641-649.

Lesson 28　Flavonoids

Flavonoids are formed in plants from the aromatic amino acids-phenylalanine and tyrosine, and malonate. The basic flavonoid structure is the **flavan nucleus**, which consists of 15 carbon atoms arranged in three rings (C_6-C_3-C_6), which are labeled A, B, and C (Figure 28.1). The various classes of flavonoids differ in the level of oxidation and pattern of substitution of the C ring, while individual compounds within a class differ in the pattern of substitution of the A and B rings. Among the many classes of flavonoids, those of particular interest to this review are flavones, flavanones, isoflavones, flavonols, flavanonols, flavan-3-ols, and anthocyanidins. Other flavonoid classes include biflavones, chalcones, aurones, and **coumarins**. Hydrolyzable tannins, proanthocyanidins (flavan-3-ol oligomers), caffeates, and lignans are all plant phenols, and they are usually classified separately.

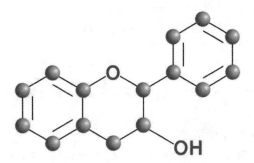

Figure 28.1　Chemical scaffold of flavonoids

Flavonoids generally occur in plants as glycosylated derivatives, and they contribute to the brilliant shades of blue, scarlet, and orange, in leaves, flowers, and fruits. Apart from various vegetables and fruits, flavonoids are found in **seeds, nuts, grains, spices**, and different medicinal plants as well in beverages, such as wine (particularly red wine), tea, and (at lower levels) beer. More specifically, the flavones apigenin and luteolin are common in **cereal grains** and aromatic herbs (parsley, rosemary, thyme), while their hydrogenated analogues hesperetin and naringin are almost exclusively present in citrus fruits. The flavonols quercetin and kaempferol are predominant in vegetables and fruits, where they are found mainly in the skin, with the exception of **onions**. Isoflavones are found most often in legumes, including soybeans, black beans, green beans, and chick peas. Alfalfa and clover sprouts and sunflower seeds also contain isoflavones. The flavan-3-ols (+)-catechin, (-)-epicatechin, (-)-epigallocatechin, and their gallate esters are widely distributed in plants, although they are very rich in tea leaves. Flavan oligomers (proanthocyanidins) are present in apples, grapes, berries, persimmon, black currant, and sorghum and barley grains. Anthocyanidins and their glycosides (anthocyanins) are natural pigments and are abundant in berries and red grape. Flavonoids play different roles in the ecology of plants. Due to their attractive colors, flavones, flavonols, and anthocyanidins may act as visual signals for pollinating insects. Because of their astringency, **catechins** and other flavanols can represent a defense system against insects harmful to the plant. Flavonoids act as catalysts in the light phase of photosynthesis and/or as regulators of iron channels involved in **phosphorylation**. They can also function as stress protectants in plant cells by scavenging ROS produced by the photosynthetic electron transport system. <u>Furthermore, because of their favorable UV absorbing properties, flavonoids protect plants from UV radiation of sun and scavenge UV-generated ROS.</u> Apart from their physiological roles in the plants, flavonoids are important components in the human diet, although they are generally considered as nonnutrients. Indeed, the level of intake of flavonoids from diet is considerably high as compared to those of vitamin C (70 mg · d^{-1}), vitamin E (7-10 mg · d^{-1}), and carotenoids (âcarotene, 2-3 mg · d^{-1}). Flavonoid intake can range between 50 and 800 mg · d^{-1}, depending on the consumption of vegetables and fruit, and of specific **beverages**, such as red wine, tea, and unfiltered beer. In particular, red wine and tea contain high levels (approximately 200 mg per glass of red wine or cup of tea) of total phenols. Thus, variations in **consumption** of these beverages are mainly responsible for the overall flavonoid intake in different national diets. Another significant source of flavonoids are different **medicinal plants** and related phytomedicines.

Selected from: *J. Nat. Prod.*, 2000, 63: 1035-1042.

Words and Expressions:

    flavonoid ['fleɪvənɒɪd] *n.* 黄酮类；类黄酮；蒲公英黄酮
    flavan nucleus 黄烷核
    coumarin ['kuːmərɪn] *n.* 香豆素
    seeds [siːdz] *n.* 种子；种子选手(seed 的复数)；种子状物；子孙后代
    nuts [nʌts] *n.* 坚果；螺母(nut 的复数)；核心
    grains [greɪns] *n.* 谷粒(grain 的复数)；双齿鱼叉，多齿鱼叉
    spices [s'paɪsɪz] *n.* 香味料，调味料(spice 的复数)
    cereal grains 粮谷；谷粒
    onions ['ʌnjəns] *n.* 洋葱(onion 的复数)
    catechin ['kætɪtʃɪn] *n.* 儿茶素；儿茶酚
    phosphorylation [fɔˌsfɔri'leɪʃən] *n.* 磷酸化作用
    beverage ['bɛvərɪdʒ] *n.* 饮料；酒水；饮料
    consumption [kən'sʌm(p)ʃ(ə)n] *n.* 消费；消耗；肺痨
    medicinal plants 药用植物

Notes:

1) Isoflavones are found most often in legumes, including soybeans, black beans, green beans, and chick peas.

异黄酮最常见于豆类，包括大豆、黑豆、青豆和鹰嘴豆。

2) Furthermore, because of their favorable UV absorbing properties, flavonoids protect plants

from UV radiation of sun and scavenge UV-generated ROS.

此外，由于黄酮类化合物具有良好的吸收紫外线的特性，它可以保护植物免受太阳的紫外线辐射，清除紫外线产生的活性氧。

Recommended Reading Materials:

1. Cushnie T P T, Lamb A J. Antimicrobial activity of flavonoids. *Int. J. Antimicrob. Ag.*, 2005, 26: 343-356.

2. Havsteen B H. The biochemistry and medical significance of the flavonoids. *Pharmacol. Therapeut.*, 2002, 96: 67-202.

Lesson 29　Modifications of Naturally Occurring Phenols

The functionalities present in agro-waste such as lignin, cardanol, tannin, palm oil and CST offer a wide variety of structural modifications to synthesize new bio-based renewable structures. The functionalities varies from **phenolic** -OH, aromatic ring, **aliphatic side chains** (saturated or unsaturated bonds), carboxylic, carbonyl groups etc. which could be either used as such or further chemically transformed. These chemicals may then acts as monomers or **oligomers** or intermediates which could be further explored for a range of polymers. Modifications of such naturally occurring phenolic compounds can be classified into three categories namely reaction due to (i) phenolic hydroxyl group (ii) aryl group (iii) side chains which are described in below section.

a) Reaction due to-OH group

The phenolic hydroxyl group can react with different structural modifiers and lead to different molecules via **nucleophilic substitution** (SNAr/SN) or condensation reaction. The structure of monomer can be tailored by using various **electrophilic centres** with different substitution groups attached, which could be further altered by using additional polymerisable sites to affect the curing process.

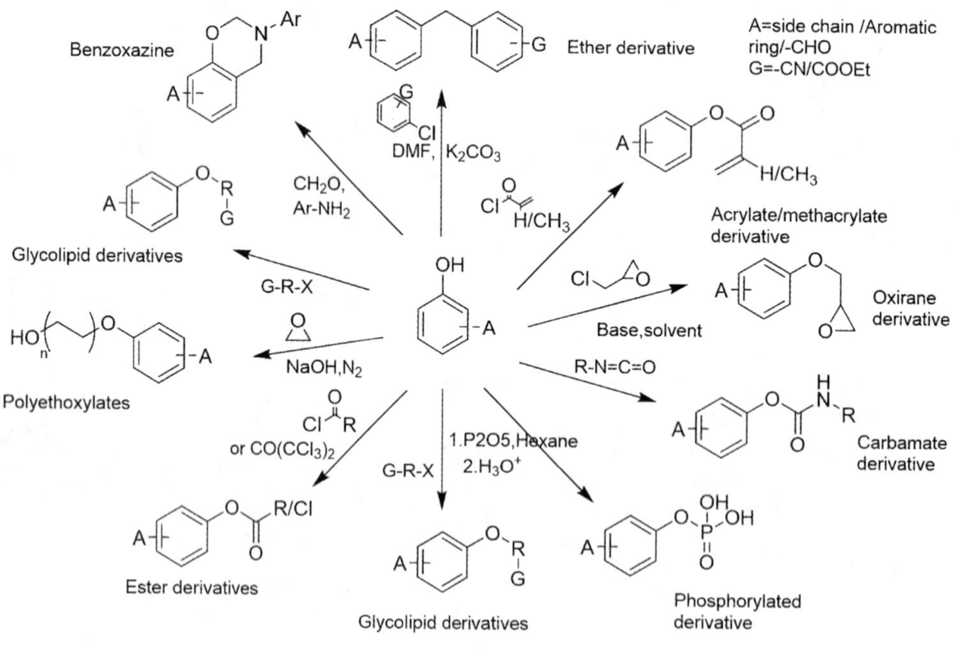

Figure 29.1　Reaction due to −OH group of phenols

b) Reaction due to-aryl group

Electrophilic aromatic substitution reaction modifies the aryl group present in cardanol, lignophenol, tannin, palm oil and CST. The main reaction studied for lignophenol, cardanol, EPFB and CST is the condensation of phenolic compound with formaldehyde. This resulted in formation of water soluble methylol derivatives of phenols (resoles) or relatively high molecular mass novolac resins. Modification of lignin prior to resin synthesis has typically been performed by reacting lignin with phenol in the presence of organic solvents such as methanol or ethanol. This process is called **phenolysis**. Initially, lignin was allowed to react with phenol before performing a condensation reaction with **formaldehyde**. Lignin was added to a mixture of phenol dissolved in ethanol, such that the lignin/phenol weight ratio was varied and known. Phenolysis of the lignin was carried out at 70 ℃ for a few hours. Organosolv Alcell lignin was also used as a replacement for phenol in PF resins. Besides Cardanol-Formaldehyde (CF) resins there are several other electrophilic reactions of cardanol were also reported such as diazotization of cardanol with different aromatic amines and **electrophilic substitution** of nitro group in benzene ring. Aryl coupling to form bisaryl derivatives lead to different set of monomers due to addition of functionality, from a monophenol to diphenol. Electropolymerisation of vanillin, eugenol to form **polyvanillin** and **polyeugenol** is also gaining importance for completely green polymers. Sulphonation of phenols extracted from the pyrolysis oil of palm shells using simultaneous sulphonation alkylation process in presence of alpha-olefin sulphonic acid may find applications as surfactant in oil fields.

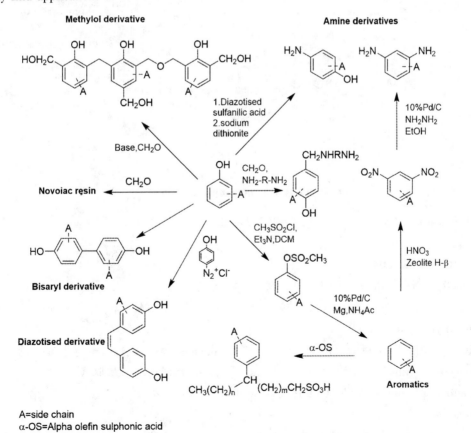

Figure 29.2 Reaction due to − aryl group of phenols

c) Reaction due to side chain

Depending on the source of phenolic compound (i. e. cardanol, palm oil, lignophenols) the side group may be alkyl (-CH_3, $C_{15}H_{31}$), alkylene (-$C_{15}H_{29}$, -$C_{15}H_{27}$, $C_{15}H_{25}$, -CH=CH-COOH), ether (-OCH_3), aldehyde group, carboxylic group etc. Cardanol double bond may be epoxidised or undergo olefin-metathesis reaction which may act as modified monomer for the formation of different set of polymers. The modification of alkylene side chain by ozonisation 445 has been carried out to introduce alcohol, aldehyde and carboxylic etc. functionalities. Ruthenium catalyzed olefin-metathesis has been successfully applied to the synthesis of biscardanol derivatives and cardanol-based porphyrins using Grubbs catalyst. Cardanol and its derivatives were used as the precursor for the synthesis of fulleropyrrolidines which could be used in medicinal chemistry, pharmaceuticals and photovoltaic applications.

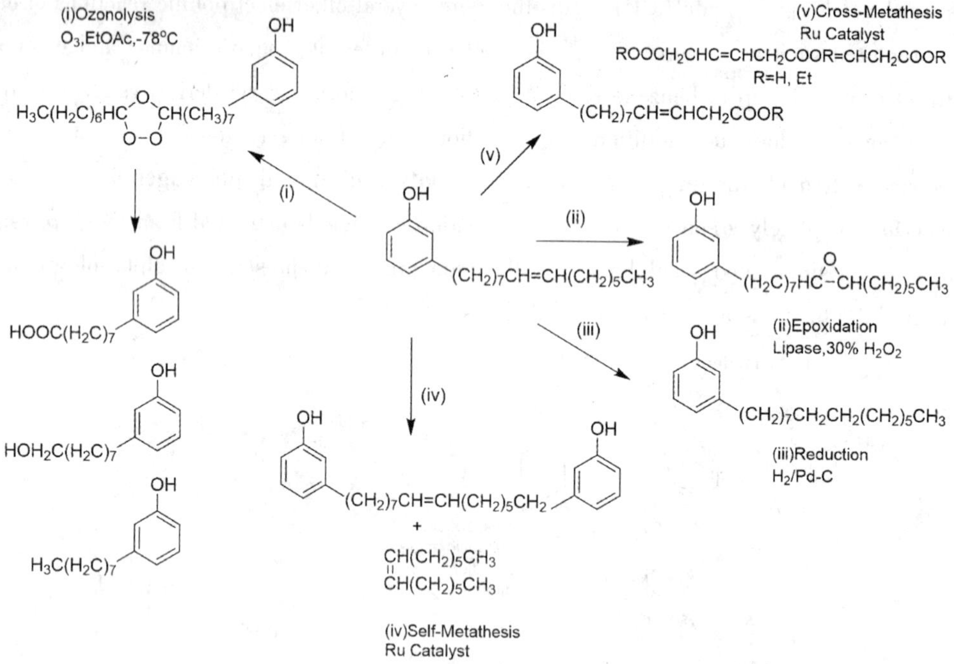

Figure 29.3 Reaction due to side chain of phenols

Selected from: *RSC Adv.*, 2014, 4: 21712 – 21752.

Words and Expressions:

Phenolic [fɪˈnɑlɪk] *adj.* [有化] 酚的，石碳酸的
aliphatic side chains [ˌælɪˈfætɪk] [saɪd] [tʃens] 脂肪侧链
carbonyl [ˈkɑrbənɪl] *n.* 羰基
oligomers [əˈlɪgəmɚs] *n.* [有化] 低聚物(oligomer 的复数)；[化学] 寡聚物
nucleophilic substitution [ˌnjuːkliəuˈfilik] [ˌsʌbstɪˈtuːʃn] 亲核取代
electrophilic centres [iˌlektrəuˈfilik] [ˈsentər] 亲电中心
phenolysis 酚解
Formaldehyde [fɔːrˈmældɪhaɪd] *n.* 蚁醛，[有化] 甲醛

electrophilic substitution　［iˌlektrəuˈfilik］［ˌsʌbstɪˈtuːʃn］　亲电取代
polyvanillin　多香兰素
polyeugenol　［pɔliˈjʊdʒəˌnɑl］　*n.* 聚丁香酚

Notes:

1) The functionalities present in agro-waste such as lignin, cardanol, tannin, palm oil and CST offer a wide variety of structural modifications to synthesize new bio-based renewable structures.

诸如木质素、鞣酸、单宁、棕榈油和 CST 之类的农业废弃物中的结构修饰可能性为合成新的生物基可再生材料(提供了条件)。

2) Lignin was added to a mixture of phenol dissolved in ethanol, such that the lignin/phenol weight ratio was varied and known.

将木质素加入溶解在乙醇中的苯酚溶液中,从而得到已知量比例的木质素/苯酚重量比。

Recommended Reading Materials:

1. Dimitrios B. Sources of natural phenolic antioxidants. *Trends in Food Science & Technology*, 2006, 17(9): 505-512.

2. Sato M, Ramarathnam N, Suzuki Y, Ohkubo T, Takeuchi M, Ochi H. Varietal differences in the phenolic content and superoxide radical scavenging potential of wines from different sources. *Journal of Agricultural and Food Chemistry*, 1996, 44(1): 37-41.

Lesson 30　Plant Oil

Today plant oils are the most important renewable raw material for the chemical industry (e. g., in Germany 30% of the 2.7 million tons of renewable raw materials in 2005 were plant oils; in total approximately 10% of all resources were renewable) and are heavily used as raw materials for surfactants, cosmetic products, and **lubricants**. In addition, plant oils have been used for decades in paint formulations, as flooring materials and for coating and resin applications. The probably best known application example is Linoleum, which was already industrially produced in 1864 and developed by F. Walton in London, UK. Its main component is linseed oil and it provides a durable and environmentally friendly alternative to, e. g., PVC floorings. Plant oils are **triglycerides** (tri-esters of glycerol with long chain fatty acids, see Figure 30.1) with varying composition of **fatty acids** depending on the plant, the crop, the season, and the growing conditions. The word 'oil' hereby refers to **triglycerides** that are liquid at room temperature. The most important parameters affecting the physical and chemical properties of such oils are the **stereochemistry** of the double bonds of the fatty acid chains, their degree of unsaturation as well as the length of the carbon chain of the fatty acids. The degree of unsaturation, which can be expressed by the iodine value (amount of iodine in g that can react with double bonds present in 100 g of sample under specified conditions) can be used as a simple parameter to divide oils into three classes:

drying (iodine value, .170; e. g. linseed oil), semi-drying (100, iodine value, 170; e. g. sunflower or soy oils) and non-drying (iodine value, 100; e. g. palm kernel oil) oils. In terms of fatty acid composition, **linseed** oil, for instance, mainly consists of linolenic (all-cis-9, 12, 15-octadecatrienoic acid) and linoleic acid (all-cis-9, 12-octadecadienoic acid), whereas in castor oil, the most abundant fatty acid is ricinoleic acid ((9Z, 12R)-12-hydroxy-9-octadecenoic acid), providing additional natural chemical functionality for modifications, cross-linking or polymerization.

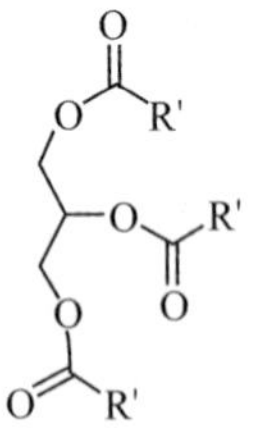

R(X, y) =	10:0	12:0	14:0	16:0	18:0	18:1	18:2	18:3	20:0
new rapeseed	–	–	0.5	4	1	60	20	9	2
sunflower	–	–	–	6	4	28	61	–	–
palm kernel	5	50	15	7	2	15	1	–	–
linseed	–	–	–	5	4	22	15	52	–
soy	–	–	–	10	5	21	53	8	0.5

Figure 30.1　Summarizes the chemical composition of some industrially important plant oils.

From Figure 30.1, it can for instance be seen that new rapeseed oil is rich in oleic acid (R = 18:1), whereas palm kernel oil is rich in lauric acid (R = 12:0). For a more complete overview and for reasons of easier reading and understanding, Figure 30.2 displays an overview of interesting fatty acids for chemical modification and the synthesis of fine chemicals, monomers and polymers. Approximately 80% of the global oil and fat production is vegetable oil, whereas 20% is of animal origin (share decreasing). About 25% is soybean, followed by **palm oil**, **rapeseed**, and sunflower oil. **Coconut** and **palm kernel oil** (laurics) contain a high percentage of saturated C12 and C14 fatty acids (compare Figure 30.1) and are most important for the production of **surfactants**. These commodity oils make highly pure fatty acids available that may be used for chemical conversions and for the synthesis of chemically pure compounds such as oleic acid (1) from "new sunflower," linoleic acid (2) from soybean, linolenic acid (3) from linseed, erucic acid (5) from rapeseed, and ricinoleic acid (9) from **castor oil**.

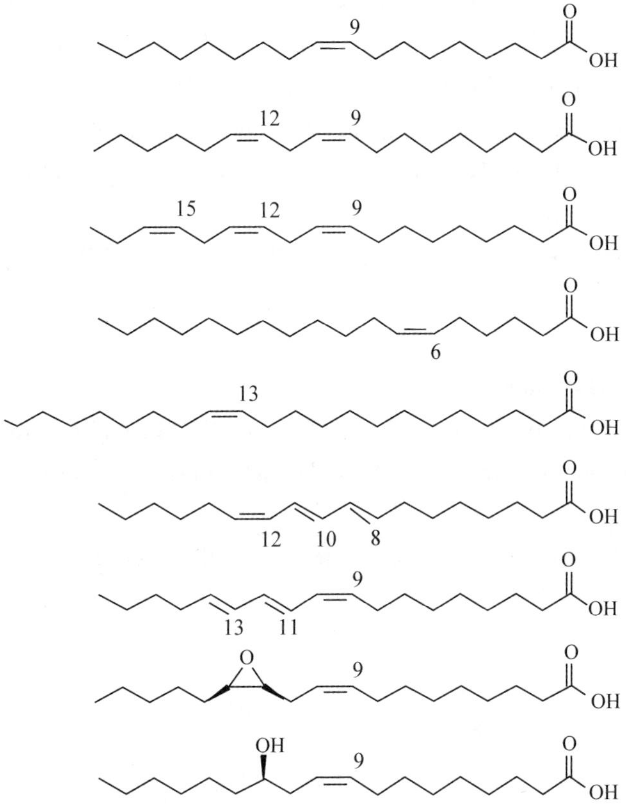

Figure 30.2 An overview of fatty acids for chemical modification and the synthesis of fine chemicals, monomers and polymers.

Selected from: *Chem. Soc. Rev.*, 2007, 36: 1788-1802.

Words and Expressions:

lubricant [ˈluːbrɪk(ə)nt] *n.* 润滑剂；润滑油；*adj.* 润滑的
triglyceride [traɪˈglɪsəraɪd] *n.* 甘油三酸酯
fatty acid 脂肪酸
stereochemistry [ˌsterɪə(ʊ)ˈkemɪstrɪ; ˌstɪərɪə(ʊ)-] *n.* 立体化学
linseed oil [ˈlɪnsiːd] *n.* 亚麻籽油；亚麻仁油
palm oil [pɑːm] *n.* 棕榈油

rapeseed ['reɪpsiːd] n. 油菜籽
coconut ['kəʊkənʌt] n. 椰子；椰子肉
palm kernel oil 棕榈仁油；棕榈坚果油
surfactant [sə'fækt(ə)nt] n. 表面活性剂；adj. 表面活性剂的
castor oil 蓖麻油

Notes:

1) The probably best known application example is Linoleum, which was already industrially produced in 1864 and developed by F. Walton in London, UK. Its main component is linseed oil and it provides a durable and environmentally friendly alternative to, e. g. , PVC floorings.

最著名的应用例子可能是油毡，它已经在1864年工业化生产，由英国伦敦的F. Walton开发。它的主要成分是亚麻油，它提供了耐用和环保的替代品，如PVC地板。

2) The most important parameters affecting the physical and chemical properties of such oils are the stereochemistry of the double bonds of the fatty acid chains, their degree of unsaturation as well as the length of the carbon chain of the fatty acids.

影响油脂理化性质的最重要参数是脂肪酸链双键的立体化学性质、不饱和程度以及脂肪酸碳链的长度。

Recommended Reading Materials:

1. Uyama H, Kuwabara M, Tsujimoto T, Nakano M, Usuki A, Kobayashi S. Green nanocomposites from renewable resources: plant oil-clay hybrid materials. *Chem. Mater.* , 2003, 15: 2492-2494.

2. Schneider M P. Plant-oil-based lubricants and hydraulic fluids. *J. Sci. Food Agric.* , 2006, 86: 1769-1780.

Lesson 31 Rosin

According to its source, **rosin** is classified into three main types: gum rosin, wood rosin, and tall oil rosin (Figure 31.1). Wood rosin is obtained from aged pine stumps, which are chipped and soaked in a solvent solution. Solid wood rosin is obtained through **distillation**. Tall oil rosin is produced during the distillation of crude tall oil, a by-product of the Kraft process of wood pulp manufacturing. Gum rosin is the most common rosin from pine resin obtained by tapping living **pine trees**. Gum rosin is the nonvolatile fraction of pine resin, while turpentine is the volatile fraction. The production of rosin is more than 1 million metric tons per year.

Figure 31.1 Image of rosin

It consists primarily of abietic- and pimaric-type rosin acids with characteristic **hydrophenanthrene** structures and about 10% neutral materials. The hydrophenanthrene structures are considered to have **cycloaliphatic** and aromatic structures, thus providing rosin with renowned hydrophobicity, which has been utilized in marine **antifouling** materials for decades by US Navy. The **predominant** rosin acid is **abietic acid** (AA) with the empirical formula $C_{20}H_{30}O_2$. Other acidic constituents of rosin differ mainly from abietic acid in that they are isomers of abietic acid having double bonds at different positions in the hydrophenanthrene moieties, which are often further hydrogenated or dehydrogenated with the aid of transition metal catalysts. The intrinsic acidity and rigidity, coupled with other chemical properties, enable rosin acids to be converted to a large number of downstream derivatives such as salts, esters, and maleic anhydride adducts, and hydrogenated, disproportionated rosins, which are used in a wide range of applications such as in the manufacture of adhesives, paper sizing agents, printing inks, solders, and fluxes, surface coatings, **insulating** materials, and **chewing gums**. It should be noted that rosin acids are a class of **stereoisomers** (3 or

4 chirality centers depending on rosin acids).

Rosin has been derivatized to containan **hydrides**, multiple-carboxyl groups, or epoxy groups as curing agents to replace some of **petroleum-derived ones** that are widely used in industry today. Exhausted coverage on thermosetting polymers involving the use of rosin is not intended in this review. We show two representative rosin derived **curing agents** to potentially replace petroleum derived 1, 2-cyclohexanedicarboxylic anhydride (CHDA) and 1, 2, 4-**benzenetricarboxylic anhydride** (BTCA). These two petroleum-based curing agents are rigid molecules due to their aromatic and cycloaliphatic structures. Rosin also has aromatic and cycloaliphatic structure. Zhang group developed new rosin-derived rigid curing agents.

Selected from: *Macromol. Rapid Commun.*, 2013, 34: 8-37.

Words and Expressions:

  rosin ['rɒzɪn] *n.* 松香；树脂；*vt.* 用松香或树脂擦抹
  distillation [ˌdɪstɪ'leɪʃn] *n.* 精馏，蒸馏，净化；蒸馏法；精华，蒸馏物
  pine [paɪn] *n.* 松树；凤梨，菠萝；*adj.* 松木的；似松的
  hydrophenanthrene 氢化菲
  cycloaliphatic [ˌsaɪkləʊˌælɪ'fætɪk] *adj.* 脂环族的
  antifouling [ˌænti'faʊlɪŋ; ˌæntaɪfaʊlɪŋ] *adj.* 防塞的，防污的；*n.* 防污漆；防污塞
  predominant [prɪ'dɒmɪnənt] *adj.* 主要的；卓越的；支配的；有力的；有影响的
  abietic acid 松香酸
  insulating ['ɪnsəletɪŋ] *adj.* 绝缘的；隔热的
  chewing gums 口香糖，泡泡糖
  stereoisomer [ˌsteriəʊ'aɪsəmə] *n.* 立体异构体
  anhydride [æn'haɪdraɪd] *n.* 酸酐；脱水物
  petroleum-derived ones 石化基衍生物
  curing agent 固化剂；硬化剂
  1, 2, 4-benzenetricarboxylic anhydride 1, 2, 4-偏苯三酸酐

Notes:

1) The hydrophenanthrene structures are considered to have cycloaliphatic and aromatic structures, thus providing rosin with renowned hydrophobicity, which has been utilized in marine antifouling materials for decades by US Navy.

该氢化菲结构被认为具有环脂和芳香结构。这一类结构使松香具有显著的疏水性，因而几十年来被美国海军用于海洋抗污材料中。

2) Other acidic constituents of rosin differ mainly from abietic acid in that they are isomers of abietic acid having double bonds at different positions in the hydrophenanthrene moieties, which are often further hydrogenated or dehydrogenated with the aid of transition metal catalysts.

松香的其他酸性成分与松香酸的主要区别在于：它们是松香酸的异构体，双键在氢化菲核的位置不同，在过渡金属催化剂的作用下，往往进一步氢化或脱氢。

Recommended Reading Materials:

1. Moustafa H, El Kissi N, Abou-Kandil A I, Abdel-Aziz M S, Dufresne A. PLA/PBAT bionanocomposites with antimicrobial natural rosin for green packaging. *ACS Appl. Mater. Interfaces*, 2017, 9: 20132-20141.

2. De Castro D O, Bras J, Gandini A, Belgacem N. Surface grafting of cellulose nanocrystals with natural antimicrobial rosin mixture using a green process. *Carbohyd. Polym.*, 2016, 137: 1-8.

Lesson 32　Extraction Technology

The history of plants being used for medicinal purpose is probably as old as the history of mankind. Extraction and **characterization** of several active phyto-compounds from these green factories have given birth to some high activity profile drugs. The potential natural anticancer drugs like **vincristine**, **vinblastine** and **taxol** can be the best example.

Recent years have shown a growing popularity and faith in the use of herbal medicine worldwide. This may be because of the realization that modern synthetic drugs have failed to provide a "cure all" guarantee to most of the human diseases with often producing undesirable side effects, which at the end turnout to be more problematic than the actual disease itself. The herbal medicine provides a ray of hope through its **cocktail** of phyto-compounds, which are believed to act in a **synergistic** manner, providing excellent healing touch with practically no undesirable side effects, provided its quality is assured off.

The modernization of herbal medicine has also raised quite a more than a few eyebrows in matters related to safety and quality of herbal medicine. In other words, the standardization and quality aspect of herbal medicine becomes a high profile issue. At present, however quality and safety related problems seems to be **overshadowing** the potential genuine benefits associated with the use of herbal medicine. The problem roots to the lack of high performance, reliable extraction, analytical techniques and methodologies for establishing a **standard therapeutic** functionality for herbal medicines. Extraction forms the first basic step in medicinal plant research because the preparation of crude extracts from plants is the starting point for the isolation and purification of chemical constituents present in plants. Yet the extraction step remains often a neglected area, which over the years has received much less attention and research. An efficient or incomplete technique means considerable constraint on the throughput of any method and involves a significant additional **workload** to staff. The traditional techniques of solvent extraction of plant materials are mostly based on the correct choice of solvents and the use of heat or/and **agitation** to increase the solubility of the desired compounds and improve the mass transfer.

Usually the traditional technique requires longer extraction time thus running a severe risk of thermal degradation for most of the phyto-constituents. The fact that one single plant can contain up to several thousand secondary **metabolites**, makes the need for the development of high performance and rapid extraction methods an absolute necessity. Keeping in pace with such requirements recent times has witnessed the use and growth of new extraction techniques with shortened extraction time, reduced solvent consumption, increased pollution prevention concern and with special care for thermolabile constituents. Novel extraction methods including microwave assisted extraction (MAE), supercritical fluid extraction (SCFE), **pressurized solvent extraction** (PSE) have drawn significant research attention in the last decade. If these techniques are explored scientifically, can prove out to be an efficient extraction technology for ensuring the quality of herbal medicines worldwide.

Selected from: *Phcog. Rev.*, 2007, 1: 7-18.

Words and Expressions:

  characterization [ˌkærəktəraɪˈzeʃən] *n.* 描述；特性描述
  vincristine [vɪnˈkrɪstiːn] *n.* 长春新碱（一种抗肿瘤药）
  vinblastine [vɪnˈblæstiːn] *n.* 长春花碱（一种抗肿瘤药）
  taxol [ˈtæksl] *n.* 紫杉醇
  cocktail [ˈkɒkteɪl] *n.* 鸡尾酒；开胃食品；*n.* 混合物；*adj.* 鸡尾酒的
  synergistic [ˌsɪnəˈdʒɪstɪk] *adj.* 协同的；协作的，协同作用的
  overshadowing *n.* 掩蔽
  standard therapeutic 标准的治疗
  workload [ˈwɜːkləʊd] *n.* 工作量
  agitation [ædʒɪˈteɪʃ(ə)n] *n.* 激动；搅动；煽动；烦乱
  metabolite [mɪˈtæbəlaɪt] *n.* 代谢物
  pressurized solvent extraction 加压溶剂萃取

Notes:

1) The herbal medicine provides a ray of hope through its cocktail of phyto-compounds, which are believed to act in a synergistic manner, providing excellent healing touch with practically no undesirable side effects, provided its quality is assured off.

具有多元组分的草药给人以一线希望。在草药质量得到保障的前提下，草药中含有的多元组分被认为具有协同作用，这种协同作用提供了良好的治愈效果，并且几乎没有不良副作用。

2) The problem roots to the lack of high performance, reliable extraction, analytical techniques and methodologies for establishing a standard therapeutic functionality for herbal medicines.

问题的根源在于缺乏高效、可靠的提取、分析技术和方法学来建立草药的标准治疗功能。

Recommended reading materials:

1. Shouqin Z, Junjie Z, Changzhen W. Novel high pressure extraction technology. *Int. J. Pharmaceut.*, 2004, 278: 471-474.

2. Pan X, Niu G, Liu H. Microwave-assisted extraction of tea polyphenols and tea caffeine from green tea leaves. *Chem. Eng. Process.*, 2003, 42: 129-133.

Lesson 33 Hydrolysis

The success of the **hydrolysis** step is essential to the effectiveness of a pretreatment operation for utilization of lignocellulose. During this reaction, the released polymer sugars, cellulose and hemicellulose are hydrolyzed into free monomer molecules readily available for fermentation conversion to bioethanol. There are two different types of hydrolysis processes that involve either acidic (sulfuric acid) or enzymatic reactions. The acidic reaction can be divided into dilute or concentrated acid hydrolysis. **Dilute** hydrolysis (1%-3%) requires a high temperature of 200-240℃ to disrupt cellulose crystals. It is followed by hexose and pentose degradation and formation of high concentrations of toxic compounds including HMF and phenolics detrimental to an effective **saccharification**. The Madison wood-sugar process was developed in the 1940s to optimize alcohol yield and reduce inhibitory and toxic byproducts.

This process uses sulfuric acid H_2SO_4 (0.5 wt%) that flows continuously to the biomass at a high temperature of 150-180℃ in a short period of time allowing for a greater sugar recovery. Concentrated acid hydrolysis, the more **prevalent** method, has been considered to be the most practical approach. Unlike dilute acid hydrolysis, concentrated acid hydrolysis is not followed by high concentrations of inhibitors and produces a high yield of free sugars (90%); however, it requires large quantities of acid as well as costly acid recycling, which makes it commercially less attractive. While acid pretreatment results in a formation of reactive substrates when acid is used as a catalyst, acid hydrolysis causes significant chemical dehydration of the monosaccharides formed such that **aldehydes** and other types of degradation products are generated. This particular issue has driven development of research to improve cellulolytic-enzymes and enzymatic hydrolysis. Effective pretreatment is fundamental to a successful enzymatic hydrolysis. During the pretreatment process, the lignocellulosic substrate enzymatic **digestibility** is improved with the increased porosity of the substrate and cellulose accessibility to cellulases. *Trichoderma reesei* is one of the most efficient and productive **fungi** used to produce industrial grade cellulolytic enzymes.

The most common cellulase groups produced by *T. reesei* that cleave the β-1,4 glycosidic bonds are **β-glucosidase**, **endoglucanases** and **exoglucanases**. However, cellulase enzymes exposed to lignin and phenolic-derived lignin are subjected to adverse effects and have demonstrated that phenolic-derived lignin have the most inhibitory effects on cellulases. This study reported that a ratio of 4 mg to 1 mg peptides, reduced by half the concentration of cellulases (i.e. β-glucosidases) from *T. reesei*. This strain was also shown to be 10 to 10 fold more sensitive to phenolics than *Aspergillus niger*. In addition to phenolic components effect on cellulases, lignin has also an adverse effect on cellulases. As mentioned previously, the lignin adverse effect has two aspects including non-productive adsorption and the limitation of the accessibility of cellulose to cellulase. Although considerable **genetic** modifications (GMs) have been deployed to transform lignin effects, lignin has been shown to be a potential source of self sustaining-energy and added-value components. Consequently, several research studies have determined practical approaches in eliminating **inhibition** of cellulases without involving GM approaches.

Selected from: *Prog. Energy Combust.*, 2012, 38: 449-467.

Words and Expressions:

hydrolysis [haɪˈdrɒlɪsɪs] *n.* 水解作用

dilute [daɪˈl(j)uːt; dɪ-] *adj.* 稀释的；淡的；*vt.* 稀释；冲淡；削弱；*vi.* 变稀薄；变淡

saccharification [səˌkærɪfɪˈkeɪʃən] *n.* 糖化(作用)

prevalent [ˈprev(ə)l(ə)nt] *adj.* 流行的；普遍的，广传的

aldehyde [ˈældɪhaɪd] *n.* 醛；乙醛

digestibility [daɪˌdʒɛstəˈbɪləti] *n.* 消化性；可消化性

fungus [ˈfʌŋɡəs] *n.* 真菌；菌类；复数 fungi [ˈfʌŋɡaɪ; ˈfʌndʒaɪ; ˈfʌndʒɪ]

β-glucosidase [gluːˈkəʊsɪdeɪs] *n.* β-葡糖苷酶

endoglucanase *n.* 内切葡聚糖酶

exoglucanase *n.* 外切葡聚糖酶

genetic [dʒɪˈnetɪk] *adj.* 遗传的；基因的；起源的

inhibition [ˌɪn(h)ɪˈbɪʃ(ə)n] *n.* 抑制；压抑；禁止

Notes:

1) While acid pretreatment results in a formation of reactive substrates when acid is used as a catalyst, acid hydrolysis causes significant chemical dehydration of the monosaccharides formed such that aldehydes and other types of degradation products are generated.

当酸作为催化剂使用时，酸预处理会导致具有反应底物的形成。与此同时，酸水解会导致单糖发生显著的化学脱水，从而产生醛和其他类型的降解产物。

2) Although considerable genetic modifications (GMs) have been deployed to transform lignin effects, lignin has been shown to be a potential source of self sustaining-energy and added-value components.

尽管大量的基因修饰(GMs)已被用于去木质化效应，但木质素已被证明是一种潜在的自我维持能源和具有较大附加值的组分。

Recommended Reading Materials:

1. Karim Z, Afrin S, Husain Q, Danish R. Necessity of enzymatic hydrolysis for production and functionalization of nanocelluloses. *Crit. Rev. Biotechnol.*, 2017, 37: 355-370.

2. Sun Y, Cheng J. Hydrolysis of lignocellulosic materials for ethanol production: a review. *Bioresource Technol.*, 2002, 83: 1-11.

Lesson 34   Hydrothermal Treatment

The term **hydrothermal** is purely of geological origin. It was first used by the British geologist Sir Roderick Murchison (1792-1871) to describe the action of water at elevated temperature and pressure, in bringing about changes in the earth's crust leading to the formation of various rocks and minerals. It is well known that the largest single crystal formed in nature (beryl crystal of >1000 kg) and some of the largest quantity of single crystals created by man in one experimental run (quartz crystals of several 1000 kg) are both of hydrothermal origin. Hydrothermal processing can be defined as any **homogeneous** (nanoparticles) or **heterogeneous** (bulk materials) reaction in the presence of aqueous solvents or **mineralizers** under high pressure and temperature conditions to dissolve and recrystallize (recover) materials that are relatively insoluble under ordinary conditions. Byrappa and Yoshimura (2001) **define** hydrothermal as any homogeneous or heterogeneous chemical reaction in the presence of a solvent (whether aqueous or non-aqueous) above the room temperature and at pressure greater than 1 atm in a closed system (Figure 34.1). However, there is still some confusion with regard to the very usage of the term hydrothermal. For example, chemists prefer to use a term, **solvothermal**, meaning any chemical reaction in the presence of a non-aqueous solvent or solvent in supercritical or near **supercritical conditions**. Similarly there are several other terms like glycothermal, alcothermal, ammonothermal, carbonothermal, lyothermal, and so on. Further, there is another school, which deals with the supercritical conditions for materials processing.

Figure 34.1   Reaction kettle for hydrothermal tratement

The supercritical solvents (water or carbon dioxide) are popularly used to carry out a wide range of chemical reactions replacing organic solvents in a number of chemical processes, including nanoparticle fabrication, extraction, chemical manufacturing, waste treatment, recycling, etc. Many researchers call this a green processing or green chemistry. Here the authors use only the term hydrothermal throughout the text to describe all the chemical reactions taking place in a closed system in the presence of a solvent, whether it is aqueous or non-aqueous, whether it is sub-critical or supercritical. It means the roots of all these diversified processes like solvothermal, ammonothermal, glycothermal or supercritical fluid technology, etc. , are hydrothermal technology only. Thus they are very closely related to one another except the solvent and the operating PT conditions. There are several methods of processing advanced materials like physical **vapour deposition**, **colloidal** chemistry approach, mechanical milling, mechanical **alloying** techniques, sol-gel, mechanical grinding, hydrothermal, **biomimitic**, flame pyrolysis, laser ablation, ultrasound techniques, electrodeposition process, **plasma** synthesis techniques, microwave techniques, other precipitation processes, etc. Among these processes, the hydrothermal technique contributes to only around 6%. However, it has been realized that the hydrothermal technique facilitates the fabrication of even the toughest or the most complex material(s) with a desired physico-chemical properties. It has several advantages over the other conventional processes like energy saving, simplicity, cost effectiveness, better nucleation control, pollution free (since the reaction is carried out in a closed system), higher dispersion, higher rate of reaction, better shape control, and lower temperature of operation in the presence of an appropriate solvent, etc. The hydrothermal technique has a lot of other advantages like it **accelerates** interactions between solid and fluid species, phase pure and homogeneous materials can be achieved, reaction kinetics can be enhanced, the hydrothermal fluids offer higher diffusivity, lower viscosity, facilitate mass transport and higher dissolving power. Most important is that the chemical environment can be suitably tailored. Although the process involves slightly a longer reaction time compared to the vapour deposition processes, or milling, it provides highly crystalline particles with a better control over its size and shape. In recent years, much attention is being paid on the hydrothermal solution chemistry through thermodynamic calculations, which facilitate the selection of a proper solvent and appropriate pressure-temperature range, which not only help in synthesizing the products, but also to control the size and shape with a significant reduction in the experimental duration. A great variety of materials like native elements, metal oxides, hydroxides, silicates, carbonates, phosphates, sulphides, tellurides, nitrides, selenides, etc. , both as particles and nanostructures like nanotubes, nanowires, nanorods, and so on have been obtained using the hydrothermal method. The method is also popular for the synthesis of a variety of forms of carbon like sp^2, sp^3 and intermediate types.

Selected from: *J. Mater. Sci.* , 2008, 43: 2085-2103.

Words and Expressions:

  hydrothermal [haɪdrə(ʊ)'θɜːm(ə)l] *adj.* 热液的；热水的

  homogeneous [ˌhɒmə(ʊ)'dʒiːnɪəs; -'dʒen-] *adj.* 均匀的；齐次的；同种的；同类的，同质的

  heterogeneous [ˌhet(ə)rə(ʊ)'dʒiːnɪəs; -'dʒen-] *adj.* 多相的；异种的；不均匀的；由不同成分形成的

  mineralizer ['mɪnərəlaɪzə] *n.* 矿化剂；造矿元素

  define [dɪ'faɪn] *vt.* 定义；使明确；规定

  solvothermal 溶剂热

supercritical condition [suːpəˈkrɪtɪk(ə)l; sjuː-] 超临界条件下
vapour deposition 气相淀积法
colloidal [kəˈlɔɪdəl] *adj.* 胶体的；胶质的；胶状的
alloying [əˈlɔɪŋ] *n.* 合金化处理；炼制合金；合铸熔合；*adj.* 合金的
biomimetic [ˌbaɪəʊmɪˈmɛtɪk] *adj.* 仿生的；生物模拟的
plasma [ˈplæzmə] *n.* 等离子体；血浆；深绿玉髓
accelerate [əkˈseləreɪt] *vt.* 使……加快；使……增速；*vi.* 加速；促进；增加

Notes:

1) Similarly there are several other terms like glycothermal, alcothermal, ammonothermal, carbonothermal, lyothermal, and so on. Further, there is another school, which deals with the supercritical conditions for materials processing.

类似地，还有一些其他的术语，如糖热、铝热、氨热、碳热、溶热等。此外，还有其他的研究机构在研究材料加工的超临界条件。

2) Here the authors use only the term hydrothermal throughout the text to describe all the chemical reactions taking place in a closed system in the presence of a solvent, whether it is aqueous or non-aqueous, whether it is sub-critical or supercritical.

在这里，在整个文本中作者只使用术语"水热"描述在一个封闭系统中有溶剂存在时发生的所有化学反应，无论溶剂是水或非水、亚临界或超临界。

3) The hydrothermal technique has a lot of other advantages like it accelerates interactions between solid and fluid species, phase pure and homogeneous materials can be achieved, reaction kinetics can be enhanced, the hydrothermal fluids offer higher diffusivity, lower viscosity, facilitate mass transport and higher dissolving power.

水热技术还具有许多其他优点，如加速了固体与流体之间的相互作用，可以获得均一相材料，可以增强反应动力学，也可以使水热流体具有较高的扩散率，较低的黏度，便于传质和较高的溶解能力。

Recommended Reading Materials:

1. Liu S, Tian J, Wang L, Zhang Y, Qin X, Luo Y, Asiri A M, Al-Youbi A O, Sun X. Hydrothermal treatment of grass: a low-cost, green route to nitrogen-doped, carbon-rich, photoluminescent polymer nanodots as an effective fluorescent sensing platform for label-free detection of Cu(II) ions. *Adv. Mater.*, 2012, 24: 2037-2041.

2. Ou H, Lo S. Review of titania nanotubes synthesized via the hydrothermal treatment: Fabrication, modification, and application. *Sep. Purif. Technol.*, 2007, 58: 179-191.

Lesson 35　Active carbon Materials

Active carbon (AC) materials are the mostly widely used **electrode** materials due to their large surface area, relatively good electrical properties and moderate cost (Figure 35.1). ACs are generally produced from physical (thermal) and/or chemical activation of various types of **carbonaceous materials** (e.g. wood, coal, nutshell, etc.) (Figure 35.1). Physical activation usually refers to the treatment of carbon precursors at high temperature (from 700 to 1200℃) in the presence of oxidizing gases such as steam, CO_2 and air. Chemical activation is usually carried out at lower temperatures (from 400℃ to 700℃) with activating agents like **phosphoric acid**, **potassium hydroxide**, sodium hydroxide and **zinc chloride**.

Depending on the activation methods as well as the carbon precursors used, ACs possessing various physicochemical properties with well developed surface areas as high as 3000 $m^2 \cdot g^{-1}$ have been produced and their electrochemical properties have been studied. It is well known that the porous structure of ACs produced by activation processes have a broad pore size distribution consisting of micropores (< 2 nm), mesopores (2-50 nm) and macropores (450 nm).

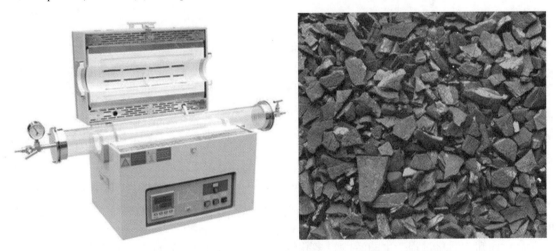

Figure 35.1　Tube furnace for preparation of activated carbon
(left) **and images of activated carbon** (right)

Several researchers have pointed out the discrepancy between the capacitance of the ACs and their specific surface area. With a high surface area up to 3000 $m^2 \cdot g^{-1}$, only a relatively small specific capacitance of 10 $mF \cdot cm^{-2}$ was obtained, much smaller than the theoretical EDL capacitance (15-25 $mF \cdot cm^{-2}$), indicating that not all pores are effective in charge accumulation. Therefore, although the specific surface area is an important parameter for the performance of EDLC, some other aspects of the carbon materials such as **pore size** distribution, pore shape and structure, electrical **conductivity** and surface functionality can also influence their electrochemical performance to a great extent. Furthermore, excessive activation will lead to large pore volume, which results in the drawbacks of low material density and conductivity. These would in turn cause a low **volumetric** energy density and loss of power capability. In addition, high active surface areas may increase the risk of decomposition of the electrolyte at the **dangling** bond positions. The presence of the acidic functionalities and moisture on the surface of ACs is responsible for the aging of the **supercapacitor** electrodes in organic **electrolytes**.

Selected from: *Chem. Soc. Rev.*, 2009, 38: 2520-2531.

Words and Expressions:

  active carbon 活性炭；活性炭黑
  electrode [ɪˈlektrəʊd] *n.* 电极；电焊条
  carbonaceous materials 富碳物质
  phosphoric acid 磷酸
  potassium hydroxide 氢氧化钾
  zinc chloride 氯化锌
  pore size 孔径；孔隙大小；气孔尺寸
  conductivity [ˌkɒndʌkˈtɪvɪtɪ] *n.* 导电性；传导性
  volumetric [ˌvɒljʊˈmetrɪk] *adj.* 体积的；容积的；测定体积的
  dangling [ˈdæŋ(ə)lɪŋ; ˈdæŋglɪŋ] *adj.* 悬挂的；摇摆的；*v.* 摇晃(dangle 的 ing 形式)
  supercapacitor *n.* 超级电容器
  electrolytes [iˈlektrəˌlaɪts] *n.* 电解质；电解质类(electrolyte 的复数形式)

Notes:

1) Physical activation usually refers to the treatment of carbon precursors at high temperature (from 700℃ to 1200℃) in the presence of oxidizing gases such as steam, CO_2 and air.

物理活化通常是指在水蒸气、二氧化碳和空气等氧化性气体环境中，在高温(700-1200 ℃)环境下，对碳前驱体的处理。

2) It is well known that the porous structure of ACs produced by activation processes have a broad pore size distribution consisting of micropores (<2 nm), mesopores (2-50 nm) and macropores (450 nm).

众所周知，活化过程产生的活性炭多孔结构具有较宽的孔径分布，包括微孔(<2 nm)、中孔(2-50 nm)和大孔(450 nm)。

Recommended Reading Materials:

1. Yang S, Han Z, Zheng F, Sun J, Qiao Z, Yang X, Li L, Li C, Song X, Cao B. $ZnFe_2O_4$ nanoparticles-cotton derived hierarchical porous active carbon fibers for high rate-capability supercapacitor electrodes. *Carbon*, 2018, 134: 15-21.

2. Zhang G, Chen Y, Chen Y, Guo H. Activated biomass carbon made from bamboo as electrode material for supercapacitors. *Mater. Res. Bull.*, 2018, 102: 391-398.

Lesson 36 Bioethanol

Lignocellulosic biomass can be transformed into bioethanol via two different **approaches**, (i. e. biochemical (Figure 36.1) or thermochemical conversion). Both routes involve degradation of the recalcitrant cell wall structure of lignocellulose into fragments of lignin, hemicellulose and cellulose. Each polysaccharide is hydrolyzed into sugars that are converted into bioethanol subsequent followed by a purification process. However, these conversion routes do not **fundamentally** follow similar techniques or pathways. The thermochemical process includes gasification of raw material at a high temperature of 800℃ followed by a catalytic reaction. Application of high levels of heat converts raw material into synthesis gas (syngas) such as hydrogen, carbon monoxide and CO_2.

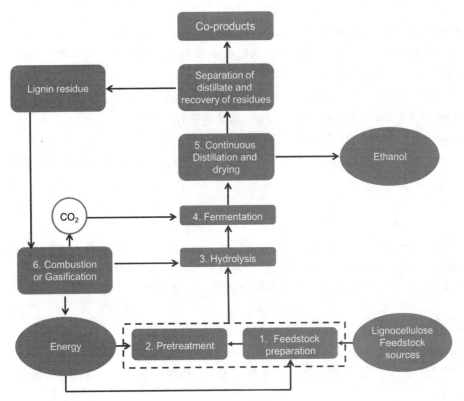

Figure 36.1 Procedure for preraration of bioethanol from lignocellulose

In the presence of catalysts, the resulting **syngas** can be utilized by the microorganism *Clostridium ljungdahlii* to form ethanol and water can be further separated by distillation. Unlike the thermochemical route, biochemical conversion involves physical (i. e. size reduction) or/and thermochemical with possible biological pretreatment. Biochemical pretreatment is mainly used to overcome **recalcitrant** material and increase surface area to optimize cellulose accessibility to cellulases. The upstream operation is followed by enzymatic or acidic hydrolysis of cellulosic materials (cellulolysis) and conversion of hemicellulose into monomeric free sugars (saccharification) subsequent to biological fermentation where sugars are fermented into ethanol and then purified via distillation. Concurrently, lignin, the most recalcitrant material of cell walls is combusted and converted into electricity and heat. Overall, **biochemical** approaches include four unit-operations namely, pretreatment, hydrolysis, **fermentation** and distillation. Currently the biochemical route is the most commonly used process.

Effective pretreatment is fundamental for optimal successful hydrolysis and downstream operations. Pretreatment **upstream** operations include mainly physical (i.e., biomass size reduction), and thermochemical processes that involve the disruption of the recalcitrant material of the biomass. This upstream operation increases substrate porosity with lignin redistribution. Therefore, it enables maximal exposure of cellulases to cellulose surface area to reach an effective hydrolysis with minimal energy consumption and a maximal sugar recovery

Selected from *Prog. Energy Combust.*, 2012, 38: 449-467.

Words and Expressions:

approach [əˈprəʊtʃ] n. 方法；途径；接近

fundamentally [ˌfʌndəˈmentəlɪ] adv. 根本地，从根本上；基础地

syngas [ˈsɪngæs] n. 合成气(指一氧化碳和氢的混合物，尤指由低级煤生产的可燃性气体，主要用于化学和生物加工以及甲醇的生产)

recalcitrant [rɪˈkælsɪtr(ə)nt] adj. 反抗的；反对的；顽强的；n. 顽抗者；不服从的人

biochemical [ˌbaɪəʊˈkemɪk(ə)l] adj. 生物化学的

fermentation [ˌfɜːmenˈteɪʃ(ə)n] n. 发酵

upstream [ˈʌpstriːm] adv. 逆流地；向上游；adj. 向上游的；逆流而上的；n. 上游部门

Notes:

1) In the presence of catalysts, the resulting syngas can be utilized by the microorganism *Clostridium ljungdahlii* to form ethanol and water can be further separated by distillation.

在催化剂的作用下，梭状芽孢杆菌利用合成气生成乙醇，通过蒸馏可以把其中的水进一步分离。

2) Pretreatment upstream operations include mainly physical (i.e., biomass size reduction), and thermochemical processes that involve the disruption of the recalcitrant material of the biomass.

预处理上游操作主要包括物理的(如生物质粉碎)和热化学过程，这些过程涉及对生物质中顽固物质的分解。

Recommended Reading Materials:

1. Zabed H, Sahu J N, Suely A, Boyce A N, Faruq G. Bioethanol production from renewable sources: current perspectives and technological progress. *Renew. Sust. Energ. Rev.*, 2017, 71: 475-501.

2. Aditiya H B, Mahlia T M I, Chong W T, Nur H, Sebayang A H. Second generation bioethanol production: a critical review. *Renew. Sust. Energ. Rev.*, 2016, 66: 631-653.

Lesson 37　Furfural

The preferred **feedstock** for **furfural** production are agricultural residues due to their homogeneity and regular availability in large quantities from food processing plants. Furfural is exclusively produced from lignocellulosic biomass by dehydrating pentose. Furfural can be produced by a one-step or a two-step process. In the one-step process, **pentosans** are hydrolyzed into xylose and then dehydrated into furfural **simultaneously**.

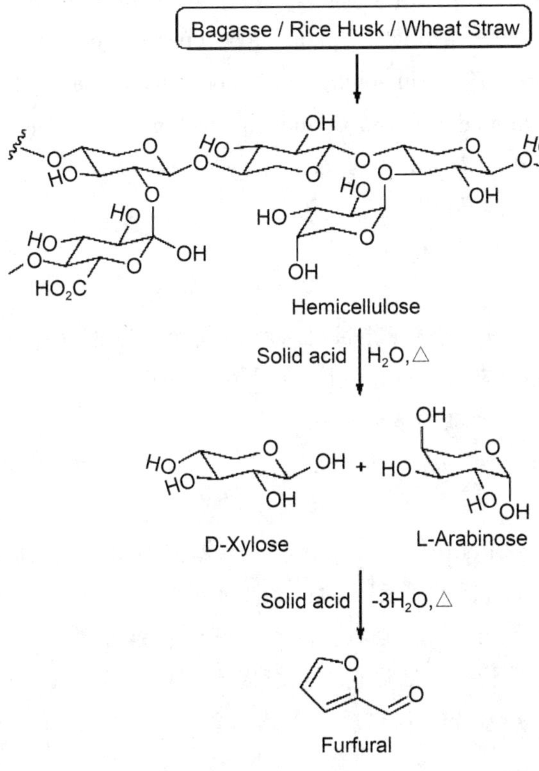

Figure 37.1　Schematic illustration of preparation of fufural from biomass

However, in the two-step process, hydrolysis of pentosans occurs under mild conditions followed by the dehydration of xylose into furfural (Figure 37.1). The advantage of the two-step process is that a higher quantity of furfural is produced when compared to the one-step process. Again, solid residues are less degraded and can be converted to other chemicals such as ethanol, phenol, glucose and others in the subsequent step by fermentation. The commercial production of furfural is by the acid hydrolysis of pentosan polysaccharides from non-food residues of food crops and wood wastes from fibrous residues of food crops. The precursors of furfural are the xylan, **arabinan** and pentosan contents of the raw biomass. The composition of these contents must be between 25%-40%. Furfural occurs naturally in many foods and is formed during the combustion of coal and wood. Humans are exposed to furfural during its production and use. Furfural is produced industrially by the use of batch or continuous reactors. In this process, the pentosan fraction of the lignocellulose is converted into **monosaccharides** (pentoses) by acid hydrolysis. When these pentoses are further dehydrated, furfural is produced. The feedstock is loaded to the digester and mixed with an aqueous solution of **inorganic** acids. The desired reaction temperature of the system is maintained by the addition of steam to the digester. In order to minimize its loss through secondary reactions of degradation and **condensation**, furfural is continuously extracted from the reactor by steam distillation in a series of distillation columns. Methanol and **acetic acid** may also be obtained as marketable byproducts, depending on the configuration of the separation process. The solid residue is separated from the liquid at the end of the reaction period and may be processed to recover the acid **catalyst**. The residual solid consists of lignin and depolymerized cellulose. This solid may be dried and burned in a **boiler** to provide steam for the reactor. Several digesters are operated in a coordinated rotation when batch reactors are used in order for the distillation plant to be operated continuously. The reaction conditions for furfural production are: 3% acid solution to lignocellulosic mass ratios of between 2:1 and 3:1, reaction temperatures of around 170-185℃ and 3h reaction time. The potential maximum furfural yields are between 45% and 50% with this technology.

Selected from: *IJAC*, 2015, 3: 42-47.

Words and Expressions:

feedstock [ˈfiːdstɒk] *n.* 原料；给料(指供送入机器或加工厂的原料)
furfural [ˈfɜːf(j)əræl] *n.* [有化] 糠醛；[有化] 呋喃甲醛
pentosan [ˈpentəsæn] *n.* 戊聚糖，多缩戊糖
simultaneously [ˌsɪmlˈteɪnɪəslɪ] *adv.* 同时地
arabinan 阿拉伯聚糖
monosaccharide [mɒnə(ʊ)ˈsækəraɪd] *n.* [有化] 单糖，单醣类(最简单的糖类)
inorganic [ɪnɔːˈɡænɪk] *adj.* [无化] 无机的；无生物的
condensation [kɒndenˈseɪʃ(ə)n] *n.* 冷凝；凝结；压缩；缩合聚合
catalyst [ˈkæt(ə)lɪst] *n.* [物化] 催化剂；刺激因素
boiler [ˈbɔɪlə] *n.* 锅炉；烧水壶，热水器；盛热水器

Notes:

1) Again, solid residues are less degraded and can be converted to other chemicals such as ethanol, phenol, glucose and others in the subsequent step by fermentation.

同样，固体残留物降解较少，可以在随后的步骤中通过发酵转化为其他化学物质，如乙醇、苯酚、葡萄糖等。

2) In order to minimize its loss through secondary reactions of degradation and condensation, furfural is continuously extracted from the reactor by steam distillation in a series of distillation columns.

为使糠醛通过降解和冷凝的二次反应损失降到最低，可在反应器中采用蒸汽蒸馏在一系

列精馏塔中连续地提取糠醛。

Recommended Reading Materials:

1. Li X, Jia P, Wang T. Furfural: a promising platform compound for sustainable production of C4 and C5 chemicals. *ACS Catal.*, 2016, 6: 7621-7640.

2. Peleteiro S, Rivas S, Alonso J L, Santos V, Parajo J C. Furfural production using ionic liquids: a review. *Bioresource Technol.*, 2016, 202: 181-191.

Lesson 38 Adhesives

Wood adhesives from renewable raw materials have been a topic of considerable interest for many years. This interest, already present in the 1940s, became more intense with the world's first oil crisis in the early 1970s but subsided as the cost of oil decreased. Since the beginning of the 21st century this interest has intensified again for a number of reasons. The main stimulus for today's renewed interest in bio-based adhesives is the acute sensitivity of the general public toward anything that has to do with the environment and its protection. It is not even this concern perse that motivates such an interest. The real concerns are rather the very strict for some synthetic adhesives almost **crippling**, government regulations which are just starting to be put into place to **allay** the environmental concerns of the public. First of all, it is necessary to define bio-based wood adhesives, or adhesives from renewable, natural, non-oil-derived raw materials. This is necessary because in its broadest sense the term might be considered to include urea-formaldehyde resins, **urea** being a non-oil-derived raw material. This, of course, is not the case.

The term "bio-based adhesive" has come to be used in a very well-specified and narrow sense to only include those materials of natural, non-mineral, origin which can be used as such or after small modifications to reproduce the behaviour and performance of synthetic resins. Thus, only a limited number of materials can currently be included in the narrowest sense of this definition. These are tannins, lignins, carbohydrates, **unsaturated oils**, liquified wood and wood welding by self adhesion. To this list in the future will surely be added proteins, blood and collagen, which already have been used to some extent for a very long time. The bio-based wood adhesives approach does not mean, however, to go back to natural products adhesives as they existed up to the 1920s and 1930s before they were **supplanted** by synthetic adhesives. The bio-based adhesives about which we are talking here are admittedly derived from natural materials, but using or requiring novel technologies, **formulations** and methods. Interest in tannins and lignins as adhesives is due to their structure, suggesting they can be used as substitutes for **phenol-formaldehyde** (PF) resins. In these cases some **formaldehyde** is still used, and in the case of lignins some other additives too. It is then necessary to **distinguish** between bio-based adhesives in which a limited amount of synthetic additives are still used, and bio-based wood adhesives where no synthetic additives are used.

Selected from: *J. Adhesion Sci. Technol.*, 2006, 20: 829-846.

Words and Expressions:

    wood adhesive 木材胶黏剂

    crippling [ˈkrɪplɪŋ] *adj.* 造成严重后果的

    allay [əˈleɪ] *vt.* 减轻；使缓和；使平静

urea [juˈriːə; ˈjuərɪə] *n.* [肥料] 尿素
unsaturated oil 不饱和油
supplant [səˈplɑːnt] *vt.* 代替；排挤掉
formulation [fɔːmjʊˈleɪʃn] *n.* 构想，规划；公式化；简洁陈述
phenol-formaldehyde 酚醛树脂
formaldehyde [fɔːˈmældɪhaɪd] *n.* 蚁醛，[有化] 甲醛
distinguish [dɪˈstɪŋgwɪʃ] *vt.* 区分；辨别；使杰出，使表现突出；*vi.* 区别，区分；辨别

Notes:

1) The main stimulus for today's renewed interest in bio-based adhesives is the acute sensitivity of the general public toward anything that has to do with the environment and its protection.

今天人们对生物性黏合剂重新产生兴趣的主要原因是公众对任何与环境及其保护有关的事物都极为敏感。

2) The term "bio-based adhesive" has come to be used in a very well-specified and narrow sense to only include those materials of natural, non-mineral, origin which can be used as such or after small modifications to reproduce the behaviour and performance of synthetic resins.

"生物性胶黏剂"一词已被用于非常明确和狭义的范围内，只包括那些天然的、非矿物的材料，这些材料可以作为生物性胶黏剂使用，或经过微小的改性复合生成的具有树脂性能产物。

Recommended Reading Materials:

1. Vnu Cec D, Kutnar A, Goršek A. Soy-based adhesives for wood-bonding - a review. *J. Adhes. Sci. Technol.*, 2017, 31: 910-931.

2. Ferdosian F, Pan Z, Gao G, Zhao B. Bio-based adhesives and evaluation for wood composites application. *Polymers*, 2017, 9: 70.

Lesson 39 Alkaloids

Alkaloids constitute a major class of natural products. The **compulsory** presence of nitrogen/(s) in their elemental contents confers basic character in these compounds save a few. Pelletier collaborated with Dumas and showed the presence of nitrogen in alkaloids. The basic or **alkali** like property of alkaloids was first observed by Sertürner while working with opium constituents and he used the term "vegetable alkali." The term alkaloid [(derived from "alkali" and Greek oeides (like)] was coined by W. Meissner, an apothecary of Halle, in 1819. Later it became a generally accepted term.

The alkaloids always enjoy an important status among the natural products because of their important biological properties guided by their **stereochemistry** and conformation, whenever applicable. These properties are manifested remarkably in some cases, e. g. , **quinine**, **reserpine**, **morphine**, **vinblastine**, **vincristine**, **camptothecine**, **taxol** (a **diterpene** alkaloid), **atropine**, **nicotine**, etc.. They are superbly remarkable drug molecules. Further, their structural **deductions** and synthesis have always been challenging. The first total synthesis of quinine by Woodward in 1945 remains a **masterpiece** in the synthetic **repository** of organic chemistry. Yet, its one stereoselective total synthesis has been achieved in 2001 after 171 years of its discovery and after 56 years of its first synthesis. This is followed by another report on catalytic **asymmetric** total synthesis of quinine in 2004 and a stereocontrolled formal synthesis of quinine and 7-hydroxyquinine by Webber *et al.* in 2008.

In almost all cases of alkaloids excepting a few, nitrogen is supplied by L-amino acids which get incorporated into the **alkaloidal** scaffold, **decarboxylation** taking place at some stage of their biosynthesis. Rarely, nitrogen is supplied by **transamination** to a carbonyl precursor (nicotine). The nitrogen(s) mostly remain(s) as a part of one or more ring(s). Sometimes nitrogen falls outside the ring and remains as a part of a side chain as amino or substituted amino group as we find in **mescaline** (a potent **hallucinogen**), **choline** (a component of vitamin B complex), **betaines** (trialkyl derivatives of amino acids, e. g. glycine betaine), **histamine**, **serotonin**, etc. In a betaine nitrogen can be a part of a ring, e. g. , **proline** betaine. They are sometimes called **protoalkaloids** or biological amines (Figure 39.1).

There are also examples of **heterocyclic** nitrogenous bases, which are biosynthetically not related to amino acids. They are known as **pseudoalkaloids**, e. g. , **caffeine**, **xanthine**, etc. (Figure 39.2).

These are conservative classifications. However, all these compounds could be generally classified as alkaloids. Though colchicine is neither basic in character nor a nitrogenous heterocycle, it has been included in the family of alkaloids as its nitrogen is supplied by an L-amino acid, L-tyrosine, and it possesses profound biological activity. Taxol (isolated from *Taxus brevifolia*) has

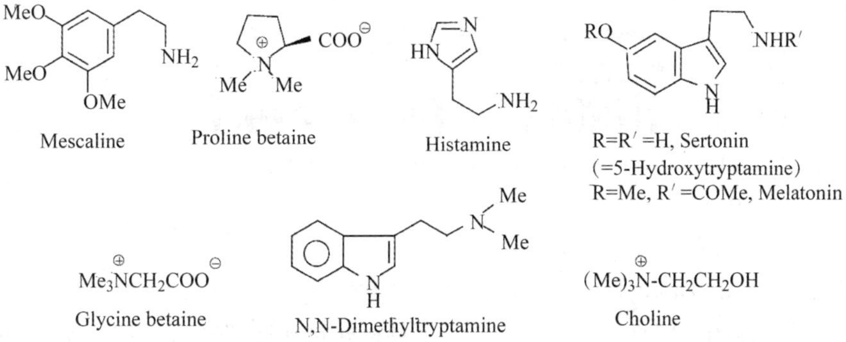

Figure 39.1 Biological amines or protoalkaloids

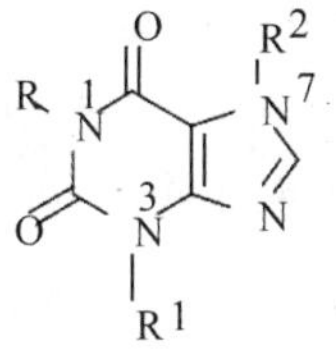

Figure 39.2 Pseudoalkaloids

R	R^1	R^2	
H	H	H	Xabthine
Me	Me	H	Theophylline(=1,3-Dimethylxanthine)
H	Me	H	Theobromine(=3,7-Dimethylxanthine)
Me	Me	Me	Caffeine(=1,3,7-Trimethylxanthine)

the gross skeleton of a diterpene (taxadiene) and a nonbasic aromatic amide side chain; it has, therefore, been included in diterpenoids. However, since its side chain is derived from rearranged L-phenylalanine, and it has exceptional biological activity, it is also included in the ambit of alkaloids by many. In fact, M. Hesse's book on "Alkaloids" carries the structure of taxol on the cover, and Wall and Wani have written an article on camptothecine and taxol in a book dedicated to alkaloids.

Alkaloids are mostly solids and are known to occur in higher plants. They are prevalent in the plants belonging to the following botanical families: Apocynaceae, Annonaceae, Amaryllidaceae, Berberidaceae, Boraginaceae, Gnetaceae, Liliaceae, Leguminoceae, Lauraceae, Loganiaceae, Magnoliaceae, Menispermaceae, Papaveraceae, Piperaceae, Rutaceae, Rubiaceae, Ranunculaceae, Solanaceae, etc. Isolated examples of families elaborating alkaloids are also known.

Selected from: S. K. Talapatra and B. Talapatra. Chapter 15: Alkaloids. General Introduction, in: *Chemistry of Plant Natural Products*. Springer-Verlag Berlin Heidelberg, 2015.

Words and Expressions:

  alkaloid　　[ˈælkəlɔɪd]　*n.* [有化] 生物碱；植物碱基
  compulsory　[kəmˈpʌlsərɪ]　*adj.* 义务的；必修的；被强制的
  alkali　　[ˈælkəlaɪ]　*n.* 碱；可溶性无机盐 *adj.* 碱性的
  stereochemistry　[ˌsterɪə(ʊ)ˈkemɪstrɪ; ˌstɪərɪə(ʊ)-]　*n.* [化学] 立体化学
  quinine　　[ˈkwɪniːn; kwɪˈniːn]　*n.* 奎宁；金鸡纳碱
  reserpine　[ˈresəpɪn]　*n.* [药] 利血平；蛇根碱
  morphine　[ˈmɔːfiːn]　*n.* [毒物][药] 吗啡
  vinblastine　[vɪnˈblæstiːn]　*n.* 长春花碱(一种抗肿瘤药)

vincristine   [vɪnˈkrɪstiːn]   n. [药]长春新碱(一种抗肿瘤药)
camptothecin   [ˌkæmptəʊˈθiːsin]   n. 喜树碱(可用以治疗癌症)
taxol   [ˈtæksl]   n. 紫杉醇(化合物)
diterpene   [daiˈtɜːpiːn]   n. [有化]双萜，[有化]二萜
atropine   [ˈætrəpiːn; -ɪn]   n. [药]阿托品(含颠茄碱)
nicotine   [ˈnɪkətiːn]   n. [有化]尼古丁；[有化]烟碱
deduction   [dɪˈdʌkʃ(ə)n]   n. 扣除，减除；推论；减除额
masterpiece   [ˈmɑːstəpiːs]   n. 杰作；绝无仅有的人
repository   [rɪˈpɒzɪt(ə)rɪ]   n. 贮藏室，仓库；知识库；智囊团
asymmetric   [ˌæsɪˈmetrɪk]   adj. 不对称的；非对称的
alkaloidal   [ˌælkəˈlɔidəl]   adj. 含碱的；生物碱的(等于 alkalogenic)
scaffold   [ˈskæfəʊld; -f(ə)ld]   n. 脚手架；vt. 给…搭脚手架；用支架支撑
decarboxylation   [ˈdiːkɑːˌbɒksɪˈleɪʃən]   n. 去碳酸基
transamination   [ˌtrænsæmɪˈneɪʃ(ə)n; ˌtrɑːns-; -nz-]   n. 转氨作用
mescaline   [ˈmeskəlɪn; -liːn]   n. [有化]墨斯卡灵(迷幻药)
hallucinogen   [həˈluːsɪnədʒ(ə)n]   n. 迷幻剂
choline   [ˈkəʊliːn; -lɪn]   n. [生化]胆碱；维生素B复合体之一
betaine   [ˈbiːteɪn]   n. [有化]甜菜碱(等于 lycine, oxyneurine, trimethylglycine)
histamine   [ˈhɪstəmiːn]   n. [生化]组胺
serotonin   [ˌserəˈtəʊnɪn]   n. [生化]血清素；5-羟色胺(血管收缩素)
proline   [ˈprəʊliːn]   n. [生化]脯氨酸
protoalkaloids   原生物碱
heterocyclic   [ˌhet(ə)rə(ʊ)ˈsaɪklɪk; -ˈsɪklɪk]   adj. 杂环的；不同环式的
pseudoalkaloid   伪生物碱
caffeine   [ˈkæfiːn]   n. [有化]/药]咖啡因；茶精(兴奋剂)
xanthine   [ˈzænθiːn]   n. 黄嘌呤；黄质

Notes：

1) The alkaloids always enjoy an important status among the natural products because of their important biological properties guided by their stereochemistry and conformation, whenever applicable.

生物碱在天然产物中一直占有重要的地位，因为只要在合适的条件下，它们的立体化学和构象决定了它们的重要生物学性质。

2) Though colchicine is neither basic in character nor a nitrogenous heterocycle, it has been included in the family of alkaloids as its nitrogen is supplied by an L-amino acid, L-tyrosine, and it possesses profound biological activity.

秋水仙碱虽然不是碱性的，也不是含氮杂环，但由于它的氮是由L-氨基酸L-酪氨酸提供的，所以它已被归入生物碱家族，并具有重要的生物活性。

Recommended Reading Materials:

Sun X, Ma J, Li C, Zang Y, Huang J, Wang X, Chen N, Chen X, Zhang D. Carbazole alkaloids with bioactivities from the stems of Clausena lansium. *Phytochem. Lett.* 2020, 38: 28-32.

Lesson 40　Biodiesel

1. Fatty-Acid Profiles of Feedstocks for Biodiesel Production

Fat is a generic term for a class of lipids in biochemistry. It refers to the greasy, solid materials found in animal tissues and in some plants. Vegetable oil is the fat extracted from plant sources, mainly from plant seeds. Chemically, plant oils and animal fats are comprised of a family of chemicals called **triglycerides**.

Triglycerides (also called **triacylglycerols**) are fatty-acid esters of glycerol: a **glycerol** combined with three fatty acids on the glycerol's hydroxyl (-OH) groups. Triglycerides are the dominate components in all naturally formed plant oils and animal fats. The chemical formula of a triglyceride is shown below (Figure 40.1), where R_1, R_2 and R_3 are longer alkyl chains. The chain lengths of the alkyl chains in naturally occurring triglycerides can be of varying length, from C10 to C30, but C16 to C18 are the most common ones. The long, nonpolar alkyl chain is an important counter balance to the polar acid functional group.

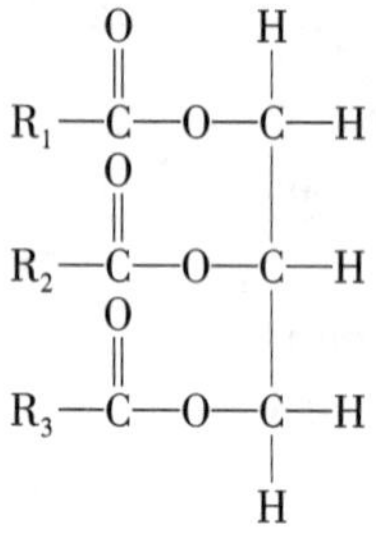

$$
\begin{array}{c}
R_1-\overset{O}{\overset{\|}{C}}-O-\overset{H}{\underset{|}{C}}-H \\
R_2-\overset{O}{\overset{\|}{C}}-O-\overset{|}{\underset{|}{C}}-H \\
R_3-\overset{O}{\overset{\|}{C}}-O-\overset{|}{\underset{H}{C}}-H
\end{array}
$$

Figure 40.1　General structure of a triglyceride

Glycerol is also known as glycerin (alternative spelling: glycerine). Glycerol is a poly hydric alcohol or triol that has three hydrophilic alcoholic hydroxyl groups (-OH). It is a colorless, odorless, **hygroscopic**, and sweet-tasting viscous liquid. In triglycerides, glycerol serves as the "backbone" to link the three fatty acids. A fatty acid is a carboxylic acid of long carbon chains. In the structure of triglycerides, three fatty acids R0-COOH, R00-COOH and R000-COOH are linked chemically to the glycerol. These three fatty acids can be the same as or different from each other. Almost without exception, only even numbers of carbon atoms are found in natural fatty acids due to the way they are biologically synthesized from acetyl Co-A, a two-carbon chemical species. Under certain conditions, fatty acids in glycerides may dissociate and become "free" fatty acids (FFAs).

Fatty acids possess different chemical and physical properties. Generally, the longer the alkyl chain of fatty acids, the higher the melting point. For fatty acids of the same length, the more double bonds a fatty acid contains, the lower its melting point. Chain length and the number of double bonds of the fatty acids are important factors affecting **biodiesel** fuel properties, which will be discussed in next section. Fatty-acid (FA) profiles of vegetable/plant oils and animal fats vary widely. Even with the same type of oil or fat, its FA profile may vary from location to location, or even from batch to batch.

2. Biodiesel Production

Biodiesel is produced by chemically reacting a vegetable oil or animal fat with a simple alcohol, such as methanol. An alkali catalyst is used. Figure 40.2 shows the global transesterification reaction for a generalized triglyceride with methanol to form methyl esters. Other alcohols such as ethanol and isopropanol can also be used as long as the resulting alkyl esters have properties that satisfy the relevant fuel standard. The industrial process to convert triglycerides to alkyl esters can be considered a competition between the transesterification reaction, which is the desired path, and **saponification**, which is undesirable. Each of these reactions will be discussed, and then the requirements to direct the reaction in the preferred direction will be presented.

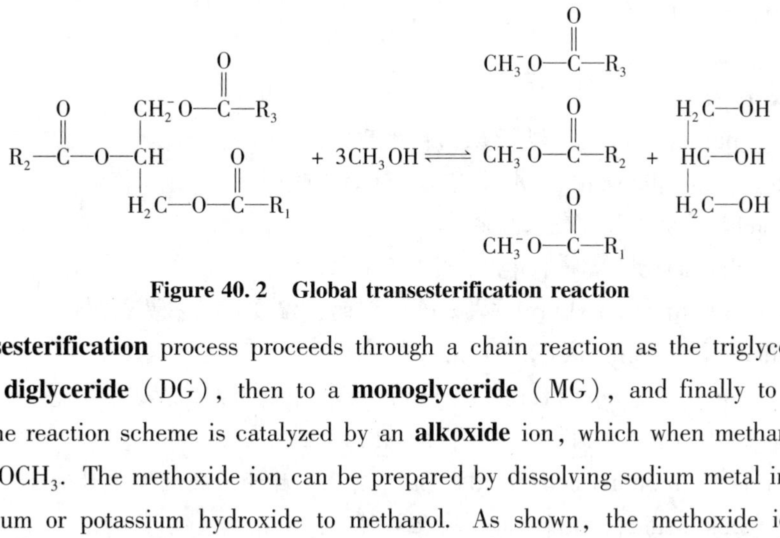

Figure 40.2 Global transesterification reaction

The **transesterification** process proceeds through a chain reaction as the triglyceride (TG) is converted to a **diglyceride** (DG), then to a **monoglyceride** (MG), and finally to free glycerin. Each step in the reaction scheme is catalyzed by an **alkoxide** ion, which when methanol is used, is **methoxide**, $^-OCH_3$. The methoxide ion can be prepared by dissolving sodium metal in methanol, or by adding sodium or potassium hydroxide to methanol. As shown, the methoxide ion attacks the triglyceride and forms a **tetrahedral** intermediate, which subsequently separates into a diglyceride and a methyl ester. The methoxide ion, which is consumed in the process, is regenerated when the diglyceride ion strips a proton from an available methanol molecule. The diglyceride and monoglyceride reactions follow similar pathways. In each case, the methoxide ion initiates the reaction, becomes part of the headgroup on the alkyl ester, and is then regenerated at the end by stripping a proton from the alcohol so the reaction can continue. Saponification is the production of salts of alkali metals with carboxylic acids, generally known as soaps. The reaction of animal fats with lye and water to produce household soap has been practiced for thousands of years. In a typical biodiesel production reactor, the transesterification and saponification reactions are occurring simultaneously. The significance of each reaction depends on the relative amounts of methoxide ion $^-OCH_3$ and hydroxide ion ^-OH in the reaction mixture. These relative amounts are determined by the following equilibrium (shown for a sodium-based catalyst):

$$CH_3OH + Na^+ + {}^-OH \rightleftharpoons Na^+ + {}^-OCH_3 + H_2O \tag{1}$$

Equation 1 clearly illustrates the process when a hydroxide catalyst is used and how it produces the necessary methoxide ions when methanol is present. Note that this reaction produces a molecule of water for each molecule of methoxide produced. Caldin and Long, in a study of the equilibrium between hydroxide and ethoxide in ethanol, found that the equilibrium strongly favors the formation of alkoxide with more than 96% of the base present as ethoxide. Assuming methanol and other alcohols behave similarly to ethanol, this equilibrium explains why hydroxides can be effective at producing the methoxide ions needed for transesterification. It also explains why hydroxides are observed to provide more soap formation. The water produced by the conversion of hydroxide to methoxide, along with water that may have been introduced with the original oil and alcohol, causes the equilibrium to favor a higher concentration of hydroxide ions and more soap production. This problem is particularly acute when free fatty acids (FFAs) are present, because the saponification of FFAs regenerates water and thus the hydroxide ions needed for additional soap production.

Selected from: Gerpen J H V, He B. Chapter 15: Biodiesel Production and Properties, in: Mark Crocker (Eds.), *Thermochemical Conversion of Biomass to Liquid Fuels and Chemicals*. Royal Society of Chemistry, 2010.

Words and Expressions:

  biodiesel　　['baɪəudiːzl]　　*n.* 生物柴油；生质柴油
  triglyceride　　[traɪ'glɪsəraɪd]　　*n.* [有化] 甘油三酸酯
  triacylglycerols　　[生化] 三酰基甘油
  glycerol　　['glɪs(ə)rɒl]　　*n.* [有化] 甘油；丙三醇
  hygroscopic　　[haɪgrə(ʊ)'skɒpɪk]　　*adj.* 吸湿的；湿度计的；易潮湿的
  saponification　　[səˌpɒnɪfɪ'keɪʃən]　　*n.* [化学] 皂化
  transesterification　　['trænsəsˌterəfɪ'keɪʃən]　　*n.* [有化] 酯基转移；酯基转移作用
  diglyceride　　[daɪ'glɪsəˌraɪd]　　*n.* [有化] 甘油二酯
  monoglyceride　　[ˌmɔnəu'glɪsəˌraid]　　*n.* [有化] 甘油一酸酯，单甘油酯
  alkoxide　　[æl'kɔksaid; -sid]　　*n.* [有化] 醇盐；酚盐
  methoxide　　[meθ'ɒksaɪd]　　*n.* [有化] 甲醇盐；[有化] 甲氧基金属
  tetrahedral　　[ˌtetrə'hiːdrəl]　　*adj.* 四面体的；有四面的

Notes:

1) For fatty acids of the same length, the more double bonds a fatty acid contains, the lower its melting point.

对于相同长度的脂肪酸，脂肪酸含有的双键越多，其熔点就越低。

2) The water produced by the conversion of hydroxide to methoxide, along with water that may have been introduced with the original oil and alcohol, causes the equilibrium to favor a higher concentration of hydroxide ions and more soap production.

氢氧化物转化为甲醇产生的水，以及可能与原始石油和酒精一起引入的水，导致平衡有

利于向产生更高浓度的氢氧化物离子和更多的肥皂方向移动。

Recommended Reading Materials:

1. Woo D G, Kim T H. Effect of kinematic viscosity variation with blended-oil biodiesel on engine performance and exhaust emission in a power tiller engine[J]. *Environmental Engineering Research*, 2020, 25(6): 946-959.

2. Lanfredi S, Matos J, da Silva S R, Djurado E, Sadouki A S, Chouaih A, Poon P S, Gonzalez E R P, Nobre M A L. K- and Cu-doped $CaTiO_3$-based nanostructured hollow spheres as alternative catalysts to produce fatty acid ethyl esters as potential biodiesel[J]. *Applied Catalysis b - envioronmental*, 2020, 272, 118986.

Lesson 41 Biomass-derived Plastic Materials

One of the most pressing issues for future generations is the development of a sustainable society. Fossil fuel used for both energy production and **plastic manufacturing** has finite availability. Within the next century, it will be nearly depleted. Approximately 7% of the global production of fossil fuel goes into the synthesis of plastic materials. In the United States alone, approximately 13% of fossil fuel consumption goes toward nonfuel chemical production. Global energy demands are expected to increase in the coming decades, further increasing the price of petroleum and other nonrenewable resources as the supply struggles to meet the **demand**. Although there is an increased investment in finding nonrenewable energy sources for global transportation and heating, the chemical industry should not be neglected. In addition to the economic influence, the undesirable environmental impact by nonrenewable resources has contributed to the **rebirth** of renewable resource alternatives. Burning fossil fuel has led to increased greenhouse gas emissions, reduced air quality, and global warming. Most plastics derived from nonrenewable resources have led to water and land pollution due to their inability to undergo biodegradation.

Figure 41.1 Biomass-derived plastic materials

The environmental concerns, along with depleting oil reserves, have led to an increased interest in the development of green plastics **derived** from renewable natural resources. The large-scale production of green plastics primarily depends on the integration of **biorefineries**. A biorefinery, as defined by the National Renewable Energy Laboratory, is a facility that integrates biomass conversion processes and equipment to produce fuels, power, and chemicals from biomass and is analogous to today's petrochemical refineries. The importance of these refineries to develop green plastics, without impacting the food and feed production in a negative manner, is **essential** for future growth.

Green plastics can be classified into three primary categories. The first class of natural resources is natural polymers including lignin, cellulose, hemicellulose, polysaccharide, and chitin. Many of these biopolymers display excellent **biocompatibility** and biodegradability. These natural polymers have long been exploited without any modifications. Currently, common approaches involve physical blending and limited chemical modifications. However, the ill-defined already-built-in macromolecular structures could be further manipulated by implementing advanced polymerization techniques and potentially serve as a building block toward diverse polymeric architectures and therefore rich properties. The second class of renewable polymers can be obtained in nature by microorganism fermentation of sugar or lipids. These polymers include polyhydroxyalkanoates (PHAs) such as **poly(hydroxybutyric acid)**. The third category of green plastics pertains to the use of small molecular biomass that can be derivatized and further polymerized. Vegetable oils, fatty acids, and lactic acids are a class of small molecular biomass, which are usually obtained directly from forestry and agriculture products or by microorganism fermentation. These materials could be precisely engineered at a molecular level into renewable polymers in a way similar to some plastics derived from petroleum chemicals. For example, poly(lactic acid) has been commercially used for over 50 years. It should be worthwhile to mention that there are increased efforts to synthesize olefins such as ethylene, propylene, and isoprene using biological routes including fermentation of biomass. As one of the major classes of petroleum chemicals, cycloaliphatic, and aromatic compounds such as benzene, cyclohexane, and cyclohexene offer rigidity and hydrophobicity to polymers derived from them. There are great opportunities to develop renewable polymers from natural resources containing rich cycloaliphatic and aromatic structures.

Selected from: *Macromolecular Rapid Communications*, 2013, 34: 8-37.

Words and Expressions:

plastic manufacturing ['plæstɪk] [mænju'fæktʃərɪŋ] 塑料制造业
demand [dɪ'mænd] n. (坚决的或困难的)要求；(顾客的)需求
rebirth [ˌriː'bɜːrθ] n. 再生；复兴
derived [dɪ'raɪvd] adj. 导出的；衍生的，派生的
biorefineries ['baɪoʊrɪ'faɪnərɪs] 生物加工
essential [ɪ'senʃl] adj. 基本的；必要的；本质的；精华的
biocompatibility [ˌbaɪokəmˌpætə'bɪlɪti] n. 生物相容性；生物适合性
poly hydroxybutyric acid ['pɑːli] [haɪˌdrɔksibjuː'tirik] ['æsɪd] 聚羟基丁酸

Notes:

1) The environmental concerns, along with depleting oil reserves, have led to an increased interest in the development of green plastics derived from renewable natural resources.

对环境问题的担忧，加上石油储量的不断减少，使得人们对开发利用可再生自然资源生产的绿色塑料越来越感兴趣。

2) Vegetable oils, fatty acids, and lactic acids are a class of small molecular biomass, which are usually obtained directly from forestry and agriculture products or by microorganism fermentation.

植物油、脂肪酸和乳酸是一类小分子生物质，通常可直接从林业、农业产品或微生物发酵中获得。

Recommended Reading Materials:

Lligadas G, Ronda J C, Galia M, Cadiz V. Renewable polymeric materials from vegetable oils: a perspective. *Materials today*, 2013, 16(9): 337-343.

Lesson 42　Biomass-derived Carbon Dots

Carbon dots (CDs) are novel fluorescent carbon nanomaterials that have sizes below 10 nm. They were first discovered **serendipitously** in 2004 as an impurity in the synthesis of single walled carbon nanotubes and were purified by preparative electrophoresis. In 2006, Sun *et al.* reported the preparation of luminescent CDs by using **laser ablation** and surface passivation. Since then, a rapidly increasing number of studies have been performed to investigate the **fascinating** properties of CDs. During the past few years, much progress has been achieved in the synthesis, properties, and applications of CDs, as reviewed by Baker *et al.*, Lee *et al.*, and Wang *et al.* Compared with traditional semiconductor quantum dots, upconverting nanoparticles, and organic dyes, photoluminescent CDs show many advantages, including high **photostability**, high **aqueous** solubility, robust chemical inertness, and easy modification. The superior biological properties of CDs, such as low toxicity and good biocompatibility, also allow potential applications in **bioimaging**, biosensors, and drug-**delivery** systems. CDs also have outstanding electronic properties as electron donors and acceptors, which endow them with wide potential in catalysis and optronics.

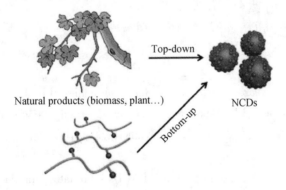

Figure 42.1　Two approachs of carbon dots synthesis by natural products

The raw carbon sources for CDs can be either man-made [e.g., candle soot, graphite, fullerene C60, ammonium citrate, glucose and ammonium hydroxide, and branched poly (ethylenimine) and ethylenediamine] or natural products (e.g., orange juice, milk, coffee grounds, *Bombyx mori* silk, green tea, egg, eggshell membrane, soy milk, flour, banana, potato, *Punica gianatum*, lotus root, pepper, honey, coriander leaves, **garlic**, Jinhua bergamot, aloe, rose flowers, limeade, hair, and rice husks). Most published reviews have focused only on the preparation and applications of CDs and ignored the raw materials. Actually, compared with man-made carbon sources, ecofriendly natural products have many advantages for the preparation of natural-product-derived carbon dots (NCDs), including cheap price and high abundance. As an additional benefit, the preparation of NCDs from natural products can convert low-value biomass waste into valuable and useful materials. Natural products that contain heteroatoms (N, S) are very suitable raw materials for the preparation of heteroatom-doped NCDs, unlike other N/S-doped CDs derived from man-made carbon sources, which require the external addition of N/S-containing compounds. New methods have been developed **specifically** for the preparation of NCDs from natural products, and these methods are different from those typically used to prepare CDs from man-made carbon sources. Although much progress has been achieved in this area, there are no reviews that specifically cover NCDs. To address this deficiency, we now provide are view of recent developments in synthetic routes, structures, properties and applications of NCDs prepared by using different natural carbon sources as raw materials.

Selected from: *ChemSusChem*, 2018, 11: 11-24.

Words and Expressions:

serendipitously [ˌserən'dɪpətəsli] *adv.* 偶然发现地，意外收获地
laser ablation ['leɪzər] [ə'bleɪʃn] 激光销蚀
fascinating ['fæsɪneɪtɪŋ] *adj.* 迷人的；吸引人的；使人神魂颠倒的
photostability ['foʊtoʊˌstə'bɪləti] *n.* 耐光性
aqueous ['eɪkwiəs] *adj.* 水的，水般的
bioimaging ['baɪoʊ'ɪmɪdʒɪŋ] 生物成像
delivery [dɪ'lɪvəri] *n.* 交付；分娩；递送
garlic ['ɡɑːrlɪk] *n.* 大蒜；蒜头
specifically [spə'sɪfɪkli] *adv.* 特别地；明确地

Notes:

1) During the past few years, much progress has been achieved in the synthesis, properties, and applications of CDs, as reviewed by Baker *et al.*, Lee *et al.*, and Wang *et al.*

在过去的几年中，Baker 等、Lee 等以及 Wang 等总结了碳点的合成、性能和应用方面的进展。

2) As an additional benefit, the preparation of NCDs from natural products can convert low-value biomass waste into valuable and useful materials.

从天然产物中制备 NCDs 的另一个好处是可以将低价值的生物质废弃物转化为有价值和有用的材料。

Recommended Reading Materials:

1. Wang Y, Hu A. Carbon quantum dots: synthesis, properties and applications. *Journal of Materials Chemistry C*, 2014, 2(34): 6921-6939.

2. Yan Z, Zhang Z, Chen J. Biomass-based carbon dots: synthesis and application in imatinib determination. *Sensors and Actuators B: Chemical*, 2016, 225: 469-473.

References

Gaoyuan Wei, 2012. Introductory chemistry speciality English[M]. 2nd Edition. Beijing: Peking University Press.

Gaoyuan Wei, 2012. Introductory chemistry speciality English[M]. 2nd Edition. Beijing: Peking University Press.

Gaoyuan Wei. Introductory chemistry speciality English [M]. 2nd Edition. Beijing: Peking University Press.

Mckendry P, 2002. Energy production from biomass (part 1): overview of biomass [J]. Bioresource Technology, 83: 37-46.

Zhang Z, Yang S, Li H, et al., 2018. Sustainable carbonaceous materials derived from biomass as metal-free electrocatalysts[J]. Advanced Materials, 31(13): 1805718.

Wang Z, Smith A T, Wang W, Sun L, 2018. Versatile nanostructures from rice husk biomass for energy applications[J]. Angewandte Chemie International Edition, 57: 13722-13734.

Hongli Zhu, Wei Luo, Peter N Ciesielski, et al., 2016. Wood-Derived Materials for Green Electronics, Biological Devices, and Energy Applications[J]. Chem Rev, 116: 9305-9374.

Jiang F, Li T, Li Y, et al., 2018. Wood-based nanotechnologies toward sustainability[J]. Advanced Materials, 30: 1703453.

Zhu M, Song J, Li T, et al., 2016. Highly anisotropic, highly transparent wood composites [J]. Advanced Materials, 28: 5181-5187.

W Schutyser, T Renders, S Van den Bosch, et al., 2018. Chemicals from lignin: an interplay of lignocellulose fractionation, depolymerisation, and upgrading [J]. Chemical Society Reviews, 47: 852-908.

Terashima N, Kitano K, Kojima M, et al., 2009. Nanostructural assembly of cellulose, hemicellulose, and lignin in the middle layer of secondary wall of ginkgo tracheid [J]. Wood Science, 55: 409-416.

Aro T, Fatehi P, 2017. Production and application of lignosulfonates and sulfonated lignin[J]. ChemSusChem, 10: 1861-1877.

Farhat W, Venditti R A, Hubbe M, et al., 2017. A review of water-resistant hemicellulose-based materials: processing and applications[J]. ChemSusChem, 10: 305-323.

Lancefield C S, Panovic I, Deuss P J, et al., 2017. Pre-treatment of lignocellulosic feedstocks using biorenewable alcohols: towards complete biomass valorisation[J]. Green Chemistry, 19: 202-214.

Rinaldi R, Jastrzebski R, Clough M T, et al., 2016. Paving the way for lignin valorisation: recent advances in bioengineering, biorefining and catalysis[J]. Angewandte Chemie International

Edition, 55: 8164 – 8215.

Abe K, Yano H, 2010. Comparison of the characteristics of cellulose microfibril aggregates isolated from fiber and parenchyma cells of Moso bamboo (phyllostachys pubescens)[J]. Cellulose, 17: 271 – 277.

Abe K, Yano H, 2009. Comparison of the characteristics of cellulose microfibril aggregates of wood, rice straw and potato tuber[J]. Cellulose, 16: 1017 – 1023.

Brandt, Agnieszka, 2013. Deconstruction of lignocellulosic biomass with ionic liquids[J]. Green Chemistry, 15: 550 – 583.

Zhang J, Wu J, Yu J, et al., 2017. Application of ionic liquids for dissolving cellulose and fabricating cellulose-based materials: state of the art and future trends[J]. Materials Chemistry Frontiers, 1(7): 1273 – 1290.

Ren Q, Wu J, Zhang J, et al., 2003. Synthesis of 1-allyl, 3-methylimidazolium-based room-temperature ionic liquid and preliminary study of its dissolving cellulose[J]. Acta Polymerica Sinica, 3: 448 – 451.

Hokkanen S, Bhatnagar A, Sillanpaa M, 2016. A review on modification methods to cellulose-based adsorbents to improve adsorption capacity[J]. Water Research, 91: 156 – 173

O'Connell D W, Birkinshaw C, O'Dwyer T F, 2008. Heavy metal adsorbents prepared from the modification of cellulose: a review[J]. Bioresource Technology, 99: 6709 – 6724.

Saito T, Isogai A, 2005. Ion-exchange behavior of carboxylate groups in fibrous cellulose oxidized by the TEMPO-mediated system[J]. Carbohydrate Polymers, 61: 183 – 190.

Dufresne A, 2013. Nanocellulose: a new ageless bionanomaterial[J]. Materials Today, 16: 220 – 227.

Elazzouzi-Hafraoui S, Nishiyama Y, Putaux J L, et al., 2008. The shape and size distribution of crystalline nanoparticles prepared by acid hydrolysis of native cellulose[J]. Biomacromolecules, 9: 57 – 65.

Filpponen I, Argyropoulos D S, 2010. Regular linking of cellulose nanocrystals via click chemistry: synthesis and formation of cellulose nanoplatelet gels[J]. Biomacromolecules, 11: 1060 – 1066.

Chunyu Chang, Lina Zhang, 2011. Cellulose-based hydrogels: present status and application prospects[J]. Carbohydrate Polymers, 84: 40 – 53.

Kelly J A, Shukaliak A M, Cheung C C, et al., 2013. Responsive photonic hydrogels based on nanocrystalline cellulose[J]. Angewandte Chemie International Edition, 52: 8912 – 8916.

Marc G, Mele G, Palmisano L, et al., 2006. Environmentally sustainable production of cellulose-based superabsorbent hydrogels[J]. Green Chemistry, 8: 439 – 444.

Chang C, He M, Zhou J, Zhang L, 2011. Swelling behaviors of pH- and salt-responsive cellulose-based hydrogels[J]. Macromolecules, 44: 1642 – 1648.

Liu H, Geng B, Chen Y, et al., 2017. Review on the aerogel-type oil sorbents derived from nanocellulose[J]. ACS Sustainable Chemistry & Engeering, 5: 49 – 66.

Mulyadi A, Zhang Z, Deng Y, 2016. Fluorine-free oil absorbents made from cellulose nanofibril aerogels[J]. ACS Applied Materials Interfaces, 8: 2732 – 2740.

Chen W, Li Q, Wang Y, et al., 2014. Comparative study of aerogels obtained from differently prepared nanocellulose fibers[J]. ChemSusChem, 7: 154-161.

Hatakeyama H, Hatakeyama T, 2010. Lignin structure, properties, and applications[J]. Advances Polymer Science, 232: 1-63.

Boeriu C G, Bravo D, Gosselink R J A, et al., 2004. Characterisation of structure-dependent functional properties of lignin with infrared spectroscopy[J]. Industrial Crops and Products, 20: 205-218.

Kai D, Tan M J, Chee P L, et al., 2016. Towards lignin-based functional materials in a sustainable world[J]. Green Chemistry, 18: 1175-1200.

Bimlesh, Lochab, Swapnil, 2014. Naturally occurring phenolic sources: monomers and polymers[J]. RSC Advanced, 4: 21712-21752.

Clifford D J, Carson D M, McKinney D E, et al., 1995. A new rapid technique for the characterization of lignin in vascular plants: thermochemolysis with tetramethylammonium hydroxide (TMAH)[J]. Organic Geochemistry, 23(2): 169-175.

Hedges J I, Ertel J R, 1982. Characterization of lignin by gas capillary chromatography of cupric oxide oxidation products[J]. Analytical Chemistry, 54(2): 174-178.

Balogun A O, Lasode O A, McDonald A G, 2014. Thermo-analytical and physico-chemical characterization of woody and non-woody biomass from an agro-ecological zone in Nigeria[J]. BioResources, 9(3): 5099-5113.

Li H, McDonald A G, 2014. Fractionation and characterization of industrial lignins[J]. Industrial Crops and Products, 62: 67-76.

Xu C, Arancon R, Labidi J, et al., 2014. Lignin depolymerisation strategies: towards valuable chemicals and fuels[J]. Chemical Society Reviews, 43: 7485-7500.

Das A, Rahimi A, Ulbrich A, et al., 2018. Lignin conversion to low-molecular-weight aromatics via an aerobic oxidation-hydrolysis sequence: comparison of different lignin sources[J]. ACS Sustainable Chemistry & Engeering, 6: 3367-3374.

Salvachúa D, Karp E M, Nimlos C T, et al., 2015. Towards lignin consolidated bioprocessing: simultaneous lignin depolymerization and product generation by bacteria[J]. Green Chemistry, 17: 4951-4967.

Changjun Liu, Huamin Wang, Ayman M Karim, 2014. Catalytic fast pyrolysis of lignocellulosic biomass[J]. Chemical Society Reviews, 43: 7594-7623.

Chu S, Subrahmanyam A V, Huber G W, 2013. The pyrolysis chemistry of a β-O-4 type oligomeric lignin model compound[J]. Green Chemistry, 15: 125-136.

Patwardhan P R, Brown R C, Shanks B H, 2011. Understanding the fast pyrolysis of lignin[J]. ChemSusChem, 4: 1629-1636.

Wenwen Zhao, Blake A Simmons, Seema Singh, et al., 2016. From Lignin Association to Nano-/Micro-particle Preparation: Extracting Higher Value of Lignin[J]. Green Chemistry, 18: 5693-5700.

Contreras S, Gaspar A R, Guerra A, et al., 2008. Propensity of lignin to associate: light scattering photometry study with native lignins[J]. Biomacromolecules, 9: 3362-3369.

Deng Y, Feng X, Zhou M, et al., 2011. Investigation of aggregation and assembly of alkali lignin using iodine as a probe[J]. Biomacromolecules, 12: 1116 – 1125.

Jia S, Cox B J, Guo X, et al., 2011. Hydrolytic cleavage of β-O-4 ether bonds of lignin model compounds in an ionic liquid with metal chlorides[J]. Industrial & Engineering Chemistry Research, 50(2): 849 – 855.

Zavrel M, Bross D, Funke M, et al., 2009. High-throughput screening for ionic liquids dissolving (ligno-) cellulose[J]. Bioresource Technology, 100.9: 2580 – 2587.

Patricia Figueiredo, Kalle Lintinen, Jouni T, et al., 2018. Properties and chemical modifications of lignin: Towards lignin – based nanomaterials for biomedical applications [J]. Progress in Materials Science, 93: 233 – 269.

Fulcrand H, Dueñas M, Salas E, et al., 2006. Phenolic reactions during winemaking and aging[J]. American Journal of Enology and Viticulture, 57(3): 289 – 297.

Gellerstedt G, Lindfors E, 1984. Structural changes in lignin during kraft cooking. Part 4. Phenolic hydroxyl groups in wood and kraft pulps[J]. Svensk Papperstidn, 87.15: R115 – R118.

Gómez-Fernández S, Ugarte L, Calvo-Correas T, et al., 2017. Properties of flexible polyurethane foams containing isocyanate functionalized kraft lignin [J]. Industrial Crops and Products, 100: 51 – 64.

Panesar S S, Jacob S, Misra M, et al., 2013. Functionalization of lignin: Fundamental studies on aqueous graft copolymerization with vinyl acetate[J]. Industrial Crops and Products, 46: 191 – 196.

Karhunen P, Rummakko P, Sipilä J, et al., 1995. The formation of dibenzodioxocin structures by oxidative coupling. A model reaction for lignin biosynthesis[J]. Tetrahedron Letters, 36(25): 4501 – 4504.

Mikulášová M, Košíková B, Alexy P, et al., 2001. Effect of blending lignin biopolymer on the biodegradability of polyolefin plastics[J]. World Journal of Microbiology and Biotechnology, 17(6): 601 – 607.

Perez J, Munoz-Dorado J, de la Rubia T, et al., 2002, Biodegradation and biological treatments of cellulose, hemicellulose and lignin: an overview[J]. International Microbiology, 5: 53 – 63.

Xiao L P, Shi Z J, Bai Y Y, et al., 2013. Biodegradation of lignocellulose by white-rot fungi: structural characterization of water-soluble hemicelluloses[J]. Bioenergy Research, 6: 1154 – 1164.

Popescu, C M, Popescu M C, Vasile C, 2010. Structural changes in biodegraded lime wood [J]. Carbohydrate Polymers, 79: 362 – 372.

Bacelo H A M, Santos, Sílvia C R, et al., 2016. Tannin-based biosorbents for environmental applications -a review[J]. Chemical Engneering Journal, 303: 575 – 587.

Arbenz A, Avérous L, 2015. Chemical modification of tannins to elaborate aromatic biobased macromolecular architectures[J]. Green Chemistry, 17: 2626 – 2646.

Khanbabaee K, van Ree T, 2001. Tannins: classification and definition[J]. Nature Product Report, 18: 641 – 649.

P G Pietta, 2000. Flavonoids as antioxidant[J]. Journal of Natural Product, 63: 1035 – 1042.

Cushnie T P T, Lamb A J, 2005. Antimicrobial activity of flavonoids[J]. International Journal of Antimicrobial Agents, 26: 343 – 356.

Havsteen B H, 2002. The biochemistry and medical significance of the flavonoids [J]. Pharmacology & Therapeutics, 96: 67 – 202.

Dimitrios B, 2006. Sources of natural phenolic antioxidants[J]. Trends in Food Science & Technology, 17(9): 505 – 512.

Sato M, Ramarathnam N, Suzuki Y, et al., 1996. Varietal differences in the phenolic content and superoxide radical scavenging potential of wines from different sources[J]. Journal of Agricultural and Food Chemistry, 44(1): 37 – 41.

Desroches M, Escouvois M, Auvergne R, et al., 2007. From Vegetable Oils to Polyurethanes: Synthetic Routes to Polyols and Main Industrial Products [J]. Chemical Society Reviews, 36: 1788 – 1802.

Uyama H, Kuwabara M, Tsujimoto T, et al., 2003. Green nanocomposites from renewable resources: plant oil – clay hybrid materials[J]. Chemistry of Materials, 15: 2492 – 2494.

Schneider M P, 2006. Plant-oil-based lubricants and hydraulic fluids [J]. Journal of the Scinence of Food and Agriculture, 86: 1769 – 1780.

Wilbon P A, Chu F, Tang C, 2013. Progress in Renewable Polymers from Natural Terpenes, Terpenoids, and Rosin[J]. Macromolecular Rapid Communications, 34: 8 – 37.

Moustafa H, El Kissi N, Abou-Kandil A I, et al., 2017. PLA/PBAT bionanocomposites with antimicrobial natural rosin for green packaging [J]. ACS Applied Materials Interfaces, 9: 20132 – 20141.

De Castro D O, Bras J, Gandini A, et al., 2016. Surface grafting of cellulose nanocrystals with natural antimicrobial rosin mixture using a green process[J]. Carbohydrate Polymers, 137: 1 – 8.

Mandel V, Mohan Y, Hemalatha S, 2007. Microwave assisted extraction-an innovative and promising extraction tool for medicinal plant research. Pharmacognosy Reviews, 1: 7 – 18.

Shouqin Z, Junjie Z, Changzhen W, 2004. Novel high pressure extraction technology[J]. International Journal of Pharmaceutics, 278: 471 – 474.

Pan X, Niu G, Liu H, 2003. Microwave-assisted extraction of tea polyphenols and tea caffeine from green tea leaves[J]. Chemical Engineering Processing, 42: 129 – 133.

Limayem A, Ricke S C, 2012. Lignocellulosic biomass for bioethanol production: Current perspectives, potential issues and future prospects[J]. Progress in Energy and Combustion Science, 38: 449 – 467.

Karim Z, Afrin S, Husain Q, et al., 2017. Necessity of enzymatic hydrolysis for production and functionalization of nanocelluloses[J]. Critical Reviews in Biotechnology, 37: 355 – 370.

Sun Y, Cheng J, 2002. Hydrolysis of lignocellulosic materials for ethanol production: a review [J]. Bioresource Technology, 83: 1 – 11.

Yoshimura M, Byrappa K, 2008. Hydrothermal processing of materials: past, present and future[J]. Journal of Materials Science, 43: 2085 – 2103.

Liu S, Tian J, Wang L, et al., 2012. Hydrothermal treatment of grass: a low-cost, green route to nitrogen-doped, carbon – rich, photoluminescent polymer nanodots as an effective fluorescent sensing platform for label-free detection of Cu(II) ions[J]. Advanced Materials, 24: 2037-2041.

Ou H, Lo S, 2007. Review of titania nanotubes synthesized via the hydrothermal treatment: Fabrication, modification, and application [J]. Separation and Purification Technology, 58: 179-191.

Zhang L L, Zhao X S, 2009. Carbon-based materials as supercapacitor electrodes [J]. Chemical Society Reviews, 38: 2520-2531.

Yang S, Han Z, Zheng F, et al., 2018. $ZnFe_2O_4$ nanoparticles-cotton derived hierarchical porous active carbon fibers for high rate-capability supercapacitor electrodes[J]. Carbon, 134: 15-21.

Zhang G, Chen Y, Chen Y, et al., 2018. Activated biomass carbon made from bamboo as electrode material for supercapacitors[J]. Materials Research Bulletin, 102: 391-398.

Zabed H, Sahu J N, Suely A, et al., 2017. Bioethanol production from renewable sources: current perspectives and technological progress[J]. Renewable & Sustainable Energy Reviews, 71: 475-501.

Aditiya H B, Mahlia T M I, et al., 2016. Second generation bioethanol production: a critical review [J]. Renewable & Sustainable Energy Reviews, 66: 631-653.

Eseyin A, Steele P, 2015. An overview of the applications of furfural and its derivatives[J]. International Journal of Advanced Chemistry, 3: 42-47.

Li X, Jia P, Wang T, 2016. Furfural: a promising platform compound for sustainable production of C4 and C5 chemicals[J]. ACS Catalysis, 6: 7621-7640.

Peleteiro S, Rivas S, Alonso J L, et al., 2016. Furfural production using ionic liquids: a review[J]. Bioresource Technology, 202: 181-191.

Pizzi A, 2006. Recent developments in eco-efficient bio-based adhesives for wood bonding: opportunities and issues[J]. Journal of Adhesion Science and Technology, 20: 829-846.

Vnučec D, Kutnar A, Goršek A, 2017. Soy-based adhesives for wood-bonding – a review [J]. Journal of Adhesion Science and Technology, 31: 910-931.

Ferdosian F, Pan Z, Gao G, et al., 2017. Bio-based adhesives and evaluation for wood composites application[J]. Polymers, 9: 70.

Talapatra S K, Talapatra B, 2015. Alkaloids. General Introduction//Chemistry of Plant Natural Products[M]. Berlin Heidelberg: Springer-Verlag.

Sun X, Ma J, Li C, et al., 2020. Carbazole alkaloids with bioactivities from the stems of Clausena lansium[J]. Phytochemistry Letters, 38: 28-32.

Gerpen J H V, He B, 2010. Biodiesel Production and Properties//Mark Crocker. Thermochemical Conversion of Biomass to Liquid Fuels and Chemicals[M]. Britain: Royal Society of Chemistry.

Woo D G, Kim T H, 2020. Effect of kinematic viscosity variation with blended-oil biodiesel on engine performance and exhaust emission in a power tiller engine[J]. Environmental Engineering Research, 25(6): 946-959.

Lanfredi S, Matos J, da Silva S R, et al., 2020. K- and Cu-doped $CaTiO_3$-based nanostructured hollow spheres as alternative catalysts to produce fatty acid ethyl esters as potential biodiesel[J]. Applied Catalysis B – Environmental, 272.

Lligadas G, Ronda J C, Galia M, et al., 2013. Renewable polymeric materials from vegetable oils: a perspective[J]. Materials today, 16(9): 337 – 343.

Zhang X, Jiang M, Niu N, et al., 2018. Natural-product-derived carbon dots: from natural products to functional materials[J]. ChemSusChem, 11: 11 – 24.

Wang Y, Hu A, 2014. Carbon quantum dots: synthesis, properties and applications[J]. Journal of Materials Chemistry C, 2(34): 6921 – 6939.

Yan Z, Zhang Z, Chen J, 2016. Biomass-based carbon dots: synthesis and application in imatinib determination[J]. Sensors and Actuators B: Chemical, 225: 469 – 473.